PRAISE FOR KIANNAH FROST

From "*The Power of Parasocial Interaction:*"

A Must-Read for Understanding Our Digital Connections. The author's exploration of attachment theory, cognitive processes, and personality traits provides valuable insights into why we form these connections. Moreover, the book doesn't shy away from addressing the dark side of parasocial interaction, making it an invaluable resource for understanding and managing these relationships in a responsible and healthy manner.

— ROCHELLE RILEY

From "*Infotainment Unleashed:*"

Infotainment Unleashed is a well-crafted book that emphasizes the importance of user interface design in creating practical and accessible infotainment experiences. A must-read for those interested in the intersection of technology and user experience.

— ALEXANDRA JIMENEZ

In Infotainment Unleashed, Frost expertly delves into the fascinating world of infotainment. This book's rich content and expert analysis make it an essential read for those interested in the impact of technology on entertainment and information.

— FRANK WILKINS

PARASOCIAL INTERACTION AND INFOTAINMENT

PARASOCIAL INTERACTION AND INFOTAINMENT

A GUIDE TO THE PSYCHOLOGY OF CELEBRITIES, INFLUENCERS, AND THE EVOLUTION OF INTEGRATED ENTERTAINMENT AND INFORMATION SYSTEMS (2-IN-1 COLLECTION)

NAVIGATING THE DIGITAL AGE

KIANNAH FROST

Book
Bound Studios

To the curious minds and the seekers of knowledge, who continuously strive to understand the intricacies of our ever-evolving digital world. This book is dedicated to those who embrace the power of technology and its potential to transform our lives, while remaining vigilant of its challenges and ethical implications.

A one-sided relationship can only run on the fuel of hope for so long before it sputters and dies, leaving behind the painful realization that love must be shared, not carried.

— BEAU TAPLIN

CONTENTS

THE POWER OF PARASOCIAL INTERACTION

EXPLORING THE PSYCHOLOGY, MEDIA, AND TECHNOLOGY BEHIND CELEBRITIES, INFLUENCERS, AND OUR UNSEEN RELATIONSHIPS

INTRODUCTION TO PARASOCIAL INTERACTION

In today's media-saturated world, we find ourselves increasingly drawn to the lives and personalities of those we see on our screens. Whether it's a charismatic talk show host, a relatable YouTuber, or a captivating fictional character, these personas have a unique ability to captivate our attention and evoke strong emotional responses. As we follow their stories, laugh at their jokes, and empathize with their struggles, we may begin to feel a sense of connection and familiarity that transcends the boundaries of traditional social relationships. This phenomenon, known as parasocial interaction, is the invisible bond between audience and persona that lies at the heart of our modern media landscape.

Parasocial interaction is a term that was first coined by psychologists Donald Horton and R. Richard Wohl in 1956 to describe the one-sided, emotionally tinged relationships that can develop between media consumers and the personas they encounter through various forms of media. While lacking the direct, reciprocal communication of typical social interactions, these relationships can nonetheless have a profound impact on our thoughts, feelings, and behaviors. As we immerse ourselves in the lives of our favorite personas, we may begin to see them as friends, role models, or even surrogate family members, shaping our values, beliefs, and aspirations in the process.

Despite the widespread prevalence and influence of parasocial interactions in our daily lives, they remain a largely misunderstood and underexplored aspect of human social behavior. This book aims to shed light on the unseen relationships that bind us to the personas we encounter in the media, delving into the psychological, sociological, and cultural factors that drive our attraction to these virtual companions. Through a combination of research, case studies, and personal anecdotes, we will explore the multifaceted world of parasocial interaction and examine both the power and perils of these invisible bonds.

As an author and industry expert, I bring an insider's perspective to the dynamics of parasocial relationships, drawing on my own experiences as both a media consumer and a creator. My goal is to provide readers with a comprehensive understanding of the complex forces that shape our parasocial interactions and to offer practical insights into how we can navigate these relationships in a healthy and constructive manner.

In conclusion, parasocial interaction is a powerful and pervasive force in our modern media landscape, shaping our emotions, identities, and social lives in profound and subtle ways. By examining the invisible bond between audience and persona, we can better understand our media consumption habits and learn to harness the potential of parasocial relationships for personal growth, connection, and fulfillment.

The Emergence of a New Social Phenomenon

Parasocial interaction refers to the one-sided relationships that individuals form with media personas, such as celebrities, fictional characters, and social media influencers. While the concept has existed for decades, the rapid advancement of technology and the rise of social media platforms have significantly amplified its prevalence and impact on society.

In the early days of mass media, parasocial interaction was primarily limited to the relationships that audiences formed with television and radio personalities. These relationships were often passive, as the audience could only observe and admire their favorite personas from afar. However, the emergence of the internet and social media has transformed the landscape of parasocial interaction, allowing for more direct and intimate connections between audiences and media figures.

Today, social media platforms like Instagram, Twitter, and YouTube enable users to follow the lives of their favorite personas in real-time, often blurring the lines between reality and fiction. These platforms provide a constant stream of content, allowing individuals to feel like they are part of the media figure's life, even though the relationship remains one-sided. This increased accessibility and intimacy have led to a surge in parasocial relationships, with many individuals forming strong emotional attachments to their favorite media personas.

The emergence of this new social phenomenon has sparked a growing interest among researchers and industry experts, as they seek to understand the psychological and sociological implications of parasocial interaction. Studies have shown that these relationships can have both positive and negative effects on individuals, ranging from increased self-esteem and a sense of belonging to feelings of loneliness and social isolation.

As the world becomes increasingly interconnected through technology, it is essential to examine the complex dynamics of parasocial interaction and its impact on society. This book aims to provide a comprehensive understanding of this fascinating phenomenon, exploring its various facets and shedding light on the unseen relationships that shape our modern world.

Shedding Light on the Unseen Relationships

In today's fast-paced and interconnected world, the lines between reality and virtuality are becoming increasingly blurred. As a result, the nature of human relationships is evolving, giving rise to new forms of social interaction. One such phenomenon is parasocial interaction which refers to the one-sided emotional bonds that individuals form with media personas. While lacking the direct reciprocity of traditional social interactions, these relationships can profoundly impact the lives of those who engage in them.

The primary purpose of this book is to shed light on these unseen relationships, delving into the complex world of parasocial interaction and exploring its various facets. By providing a comprehensive understanding of this phenomenon, the book aims to bridge the gap between academic research and the experiences of individuals who engage in parasocial relationships.

As an author and industry expert, I have witnessed firsthand the power and influence that parasocial interactions can have on individuals and society as a whole. Through this book, I hope to provide readers with a better understanding of the psychological, emotional, and social implications of these relationships. By examining the factors that contribute to the development of parasocial bonds, the book will also offer insights into the ways in which these connections can be both beneficial and detrimental to those who engage in them.

Furthermore, this book seeks to explore the broader implications of parasocial interaction in the context of the ever-evolving media landscape. As technology continues to advance and reshape the ways in which we consume and engage with media, it is crucial to understand the potential consequences of these changes on our social lives. By examining the role of parasocial interaction in the age of social media, streaming platforms, and virtual reality, this book will provide a timely and relevant analysis of the ways in which these new forms of media are shaping our understanding of human connection.

In conclusion, the purpose of this book is to shed light on the unseen relationships that exist between individuals and media personas, providing a comprehensive exploration of the multifaceted world of parasocial interaction. By offering an insider's perspective on the dynamics of these relationships, the book aims to provide readers with a deeper understanding of the power and perils of parasocial interaction, ultimately fostering a more informed and nuanced conversation about this increasingly prevalent social phenomenon.

An Insider's View on the Dynamics of Parasocial Relationships

As an author and industry expert, I have had the unique opportunity to delve deep into the world of parasocial interaction, both as an observer and as a participant. Over the years, I have witnessed firsthand the powerful impact that these invisible bonds can have on individuals and society as a whole. In this section, I will share my personal insights and experiences, shedding light on the complex dynamics that underpin parasocial relationships.

One of the most striking aspects of parasocial interaction is the sheer intensity of the emotional connections that can form between audience members and media personas. I have seen fans who are fiercely loyal to their favorite celebrities, defending them against any perceived slight or criticism. I have also witnessed the profound sense of loss and grief that can accompany the death or downfall of a beloved public figure. These emotional responses are not unlike those experienced in real-life relationships, highlighting the power of parasocial bonds.

Another fascinating aspect of parasocial interaction is the way in which it can blur the lines between reality and fantasy. In my research, I have encountered individuals who have become so deeply invested in the lives of their favorite characters or celebrities that they struggle to separate fact from fiction. This can lead to a distorted perception of reality and a tendency to prioritize the needs and desires of the media persona over one's own well-being.

Despite the potential pitfalls of parasocial interaction, I have also seen the positive impact these relationships can have on individuals. For many people, engaging with media personas provides a sense of comfort, companionship, and belonging. In some cases, parasocial relationships can even catalyze personal growth and self-improvement, as individuals are inspired by the qualities and achievements of their favorite celebrities.

As an industry expert, I have also worked with media professionals who are acutely aware of the power of parasocial interaction. Many of these individuals strive to use their platform responsibly, fostering positive connections with their audience and promoting messages of empathy, understanding, and social responsibility.

In conclusion, my insider's perspective on the dynamics of parasocial relationships has revealed a complex and multifaceted phenomenon that is deeply ingrained in our modern media landscape. As we continue to explore the world of parasocial interaction, we must recognize both the power and the perils of these invisible bonds and strive to navigate them with wisdom and compassion.

The Power and Perils of Parasocial Interaction

As we reach the conclusion of this introductory chapter, it is essential to recognize both the power and the perils that parasocial interaction holds in our increasingly interconnected world. The invisible bond between audience and persona has the potential to transform the way we consume media, engage with public figures, and even perceive ourselves. However, it is crucial to approach this phenomenon with a balanced perspective, acknowledging the potential pitfalls and challenges that may arise from these one-sided relationships.

On the one hand, parasocial interaction can be a powerful force for good. It allows individuals to feel connected to others, even when they are physically distant or separated by cultural barriers. This sense of connection can foster empathy, understanding, and a sense of belonging, which are all essential components of a healthy and thriving society. Furthermore, parasocial relationships can inspire individuals to pursue their passions, learn new skills, and engage in meaningful conversations about important issues. In this way, the power of parasocial interaction lies in its ability to bring people together and create a sense of shared experience.

However, the perils of parasocial interaction must also be acknowledged and addressed. One of the primary concerns is the potential for individuals to become overly invested in their parasocial relationships to the detriment of their real-life connections and well-being. This can lead to isolation, loneliness, and even depression, as individuals struggle to reconcile the disparity between their imagined relationships and their actual social lives. Additionally, parasocial interaction can sometimes foster unhealthy or unrealistic expectations, as individuals may idealize public figures and project their own desires and fantasies onto them. This can result in disappointment, disillusionment, and even feelings of betrayal when the persona inevitably fails to live up to these expectations.

As an author and industry expert, I aim to provide readers with a comprehensive understanding of the complex world of parasocial interaction while offering practical guidance on navigating its potential pitfalls. Throughout this book, we will delve deeper into the various facets of parasocial relationships, exploring topics such as the psychological underpinnings of attachment, the role of social media in fostering parasocial connections, and the ethical considerations surrounding the creation and consumption of persona-driven content. By shedding light on the unseen relationships that shape our media landscape, I hope to empower readers with the knowledge and tools they need to engage with parasocial interaction in a healthy, balanced, and informed manner.

In the end, the power and perils of parasocial interaction are two sides of the same coin. By recognizing and understanding both aspects, we can harness the potential of this social phenomenon to enrich our lives and foster a more connected, empathetic, and engaged society.

1

THE HISTORY AND EVOLUTION OF PARASOCIAL INTERACTION

A breathtaking scene of Dionysus, the god of wine and theater, descending from Mount Olympus to grace the stage, surrounded by a chorus of ecstatic revelers, with grapevines and ivy leaves swirling around them in a mesmerizing dance.

Parasocial interaction, a term coined by Donald Horton and R. Richard Wohl in 1956, refers to the one-sided relationships that individuals form with media figures, such as celebrities, fictional characters, and social media influencers. While lacking direct interaction and reciprocity, these relationships can still evoke strong emotional connections and attachments. To better understand the phenomenon of parasocial interaction, it is essential to delve into its historical context and trace its evolution over time.

The concept of parasocial interaction has its roots in the early days of human communication and storytelling. Long before the advent of mass media, people were captivated by stories and characters, whether through oral traditions, plays, or literature. These early instances of parasocial phenomena demonstrate the innate human desire for connection and the power of stories to foster emotional bonds, even with non-existent or distant figures.

With the emergence of mass media in the 20th century, the potential for parasocial interaction grew exponentially. Radio, television, and film allowed audiences to develop relationships with a vast array of characters and personalities, often on a daily basis. As media consumption became an integral part of modern life, the line between reality and fiction began to blur, leading to more complex and nuanced parasocial relationships.

The rise of the internet and social media platforms in the 21st century has further transformed the landscape of parasocial interaction. Today, individuals can consume media and actively participate in creating and disseminating content. This shift has given rise to a new breed of media figures – social media influencers – who often share intimate details of their lives with their followers, fostering a sense of familiarity and closeness that can rival or even surpass, real-life relationships.

As we examine the history and evolution of parasocial interaction, it becomes evident that this phenomenon is not merely a byproduct of modern technology but a deeply ingrained aspect of human nature. By exploring the various factors that have shaped parasocial relationships over time, we can gain valuable insights into the role these connections play in our lives and their potential consequences on society and culture. In the following sections, we will delve deeper into the early instances of parasocial phenomena, the impact of mass media, and the role of technological advancements in shaping the future of parasocial interaction.

Early Instances and Recognition of Parasocial Phenomena

This section will delve into the early instances and recognition of parasocial phenomena, tracing its roots and exploring how it has been acknowledged and studied throughout history. By understanding the origins of parasocial interaction, we can better appreciate its significance and impact on our modern-day society.

Parasocial interaction, as a concept, was first introduced by psychologists Donald Horton and R. Richard Wohl in 1956. However, the phenomenon has likely existed for centuries, as humans have always been drawn to forming connections with others, even without direct interaction. Early instances of parasocial phenomena can be found in various forms of art, literature, and entertainment, where audiences developed emotional attachments to fictional characters or admired public figures from afar.

One of the earliest examples of parasocial interaction can be traced back to ancient Greek theater, where audiences would become emotionally invested in the lives and fates of the characters portrayed on stage. This emotional attachment was further amplified by the use of masks, which allowed actors to embody multiple roles and create a sense of familiarity and intimacy with the audience.

As time progressed, the emergence of novels and serialized literature in the 18th and 19th centuries provided new opportunities for readers to form parasocial relationships with fictional characters. The serialized format allowed readers to follow the lives of these characters over extended periods, fostering a sense of attachment and emotional investment. This phenomenon was particularly evident in the works of

Charles Dickens, whose vivid and relatable characters captured the hearts of readers across the globe.

The advent of mass media in the 20th century, particularly radio and television, marked a significant turning point in the evolution of parasocial interaction. Radio personalities and television characters became household names as audiences welcomed them into their homes and daily routines. This newfound intimacy between audience and performer laid the groundwork for the study of parasocial phenomena as psychologists and sociologists began to explore the implications of these one-sided relationships.

Horton and Wohl's seminal 1956 paper, "Mass Communication and Para-social Interaction: Observations on Intimacy at a Distance," was the first to formally define and analyze the concept of parasocial interaction. They described it as a "seeming face-to-face relationship between spectator and performer," emphasizing the illusion of intimacy and reciprocity that characterizes these interactions. This pioneering work laid the foundation for decades of research on the psychological, social, and cultural implications of parasocial relationships.

In summary, the early instances and recognition of parasocial phenomena can be traced back to ancient Greek theater, evolving through literature and reaching new heights with the advent of mass media. The formal study of parasocial interaction began with Horton and Wohl's groundbreaking research in the 1950s, paving the way for a deeper understanding of the complex dynamics between audiences and the figures they admire from afar. As we continue to explore the history and evolution of parasocial interaction, we will see how technological advancements and the emergence of social media have further transformed the landscape of these unique relationships.

The Role of Mass Media in Shaping Parasocial Relationships

The advent of mass media has played a pivotal role in developing and proliferating parasocial relationships. As a powerful communication tool, mass media can reach and influence a vast audience, transcending geographical boundaries and cultural differences. This section delves into the ways mass media has shaped parasocial interactions, from the early days of radio and television to the present-day digital landscape.

In the early 20th century, radio emerged as the first mass medium that allowed audiences to connect with personalities and characters on a regular basis. Radio programs, such as serialized dramas and talk shows, provided listeners with a sense of familiarity and intimacy with the hosts and characters, fostering the formation of parasocial bonds. The unique aspect of radio was its ability to create vivid mental images in the minds of listeners, allowing them to feel a personal connection with the voices they heard.

The introduction of television in the mid-20th century further amplified the impact of mass media on parasocial relationships. Television provided a visual element that radio lacked, enabling viewers to see and observe the mannerisms, expressions, and body language of the personalities and characters they followed. This visual component intensified the sense of connection and familiarity, as audiences could now put a face to the voice they had previously only heard. Popular television programs, such as sitcoms, soap operas, and talk shows, became platforms for viewers to develop parasocial attachments to the characters and hosts.

As mass media continued to evolve, so too did the nature of parasocial relationships. The emergence of reality television in the late 20th century blurred the lines between fiction and reality, as viewers were invited into the personal lives of real people. This genre of television fostered a new level of parasocial interaction, as audiences felt a heightened sense of connection to the individuals they watched on screen. The perceived authenticity of reality television further strengthened the emotional bonds between viewers and the personalities they followed.

The 21st century has witnessed a digital revolution, with the internet and social media platforms becoming integral components of modern mass media. Social media, in particular, has had a profound impact on the formation and maintenance of parasocial relationships. Platforms such as Facebook, Twitter, Instagram, and YouTube have given audiences unprecedented access to the lives of celebrities, influencers, and content creators. This direct and interactive communication has fostered a sense of intimacy and familiarity that surpasses traditional mass media channels.

In conclusion, mass media has played a crucial role in shaping parasocial relationships throughout history. From the early days of radio and television to the current digital era, mass media has provided the means for audiences to connect with personalities and characters in a way that transcends physical proximity. As technology continues to advance, the nature of parasocial interactions will likely evolve, further blurring the lines between reality and fiction, and deepening the emotional bonds between audiences and the figures they follow.

Technological Advancements and the Emergence of Social Media

The rapid development of technology and social media's emergence have significantly influenced how we experience parasocial interaction today. As we delve into this section, we will explore how these advancements have shaped the nature of parasocial relationships and their implications on our daily lives.

The invention of the television in the 20th century marked a turning point in the history of parasocial interaction. It allowed audiences to form connections with characters and personalities on a more intimate level, as they could now see and hear them in their living rooms. This visual and auditory access to public figures created a sense of familiarity and closeness, which further intensified parasocial bonds.

However, it was the advent of the internet and social media that truly revolutionized parasocial interaction. Platforms such as Facebook, Twitter, Instagram, and YouTube enable celebrities and influencers to share their personal lives with their followers in real time. This unprecedented level of access and interaction has blurred the lines between public and private spheres, making it easier for fans to feel like they are part of their favorite personalities' lives.

Social media has also provided a space for fans to engage with one another, creating communities centered around shared interests and admiration for specific public figures. This sense of belonging and camaraderie can further strengthen parasocial relationships, as fans feel more connected to the object of their affection through the support and validation of their peers.

Moreover, the interactive nature of social media has allowed public figures to engage directly with their audience, responding to comments and messages, and even forming friendships with some of their most dedicated followers. This two-way communication has given fans the illusion of a genuine, personal connection with their favorite personalities, further fueling the intensity of parasocial relationships.

While the rise of social media has undoubtedly contributed to the growth and prevalence of parasocial interaction, it has also raised concerns about the potential negative effects on individuals and society as a whole. Issues such as addiction, self-esteem, and the erosion of genuine interpersonal relationships have been linked to excessive social media consumption and the pursuit of parasocial connections.

In conclusion, technological advancements and the emergence of social media have significantly impacted the way we experience and engage in parasocial interaction. As we continue to navigate this digital landscape, it is crucial to remain aware of the potential consequences of these virtual connections and strive to maintain a healthy balance between our online and offline lives.

The Impact of Parasocial Interaction on Society and Culture

Parasocial interaction has profoundly impacted society and culture, shaping how we perceive and engage with public figures, celebrities, and fictional characters. This section will explore the various ways in which parasocial interaction has influenced our social lives, mental health, and cultural norms.

One of the most significant effects of parasocial interaction is the blurring of boundaries between public and private lives. As people form emotional attachments to media figures, they often feel entitled to know intimate details about their lives. This has led to an increased demand for personal information, which in turn has fueled the growth of tabloid journalism, gossip websites, and reality television. The constant exposure to the private lives of public figures has normalized voyeurism and created a culture of surveillance, where privacy is often sacrificed for the sake of entertainment.

Parasocial interaction has also affected the way we form and maintain relationships in our own lives. As we become more invested in the lives of media figures, we may begin to prioritize these parasocial relationships over our real-life connections. This can lead to feelings of isolation and loneliness and a decreased ability to form deep, meaningful relationships with others. Additionally, the idealized portrayals of relationships in the media can create unrealistic expectations for our own romantic and platonic connections, leading to dissatisfaction and disappointment.

The impact of parasocial interaction on mental health cannot be overlooked. For some individuals, the emotional investment in media figures can lead to feelings of envy, inadequacy, and low self-esteem. Comparing oneself to the seemingly perfect lives of celebrities and influencers can exacerbate feelings of worthlessness and contribute to the development of mental health issues such as depression and anxiety. On the other hand, parasocial relationships can also provide a sense of comfort and companionship for those who may struggle with social anxiety or have difficulty forming connections in their everyday lives.

Culturally, parasocial interaction has given rise to the phenomenon of celebrity worship, where fans idolize and obsess over their favorite media figures. This can manifest in various ways, from collecting memorabilia and attending fan conventions to engaging in more extreme behaviors such as stalking and harassment. The idolization of celebrities has also led to the commodification of fame, where individuals are encouraged to pursue fame for its own sake rather than for their talents or accomplishments.

In conclusion, parasocial interaction has significantly impacted society and culture, shaping how we perceive and engage with public figures and influencing our relationships and mental health. As we continue to navigate the ever-evolving landscape of media and technology, we must remain aware of the potential consequences of parasocial interaction and strive to maintain a healthy balance between our real-life connections and our engagement with media figures.

Reflecting on the Past and Envisioning the Future of Parasocial Interaction

As we reach the conclusion of this chapter, it is essential to take a moment to reflect on the historical journey of parasocial interaction and consider the potential future developments in this fascinating area of study. From its early recognition in the 1950s to the present day, parasocial interaction has evolved and expanded, driven by technological advancements and the ever-changing landscape of mass media.

Throughout history, parasocial phenomena have been observed in various forms, such as the emotional attachments to fictional characters in literature and the admiration of public figures. However, the emergence of mass media, particularly television, truly brought parasocial relationships to the forefront of our collective consciousness. As audiences began to connect with TV personalities, researchers explored the psychological underpinnings of these one-sided relationships.

The internet and social media platforms have further transformed the nature of parasocial interaction. Today, we can engage with celebrities, influencers, and content creators in ways that were once unimaginable. The lines between reality and fiction have become increasingly blurred, as individuals curate their online personas and share intimate details of their lives with millions of followers. This unprecedented level of access has led to a new era of parasocial relationships characterized by a sense of intimacy and connection that rivals those of real-life friendships.

As we look to the future, it is crucial to consider the potential implications of parasocial interaction on society and culture. On the one hand, these relationships can provide a sense of comfort and companionship, particularly for those who may struggle to form connections in their offline lives. Additionally, parasocial interaction can foster a sense of community and belonging as fans unite over shared interests and passions.

On the other hand, there are potential negative consequences to consider. The increasing prevalence of parasocial relationships may contribute to a decline in face-to-face social interaction, as individuals turn to their screens for connection and validation. Furthermore, the curated nature of online personas can lead to unrealistic expectations and comparisons, potentially impacting self-esteem and mental health.

As we continue to navigate the digital age, it is essential for researchers, mental health professionals, and society as a whole to remain vigilant and adaptive in understanding the complexities of parasocial interaction. By doing so, we can work towards harnessing the positive aspects of these relationships while mitigating the potential negative consequences.

In conclusion, the history and evolution of parasocial interaction serve as a testament to the human desire for connection and the power of mass media in shaping our social experiences. As we move forward, it is our responsibility to continue exploring this fascinating phenomenon, ensuring that we are equipped to navigate the ever-changing landscape of human interaction in the digital age.

Chapter Summary

- Parasocial interaction refers to the one-sided relationships that individuals form with media figures, such as celebrities, fictional characters, and social media influencers. This phenomenon has its roots in early human communication and storytelling.
- Early instances of parasocial phenomena can be traced back to ancient Greek theater, evolving through literature and reaching new heights with the advent of mass media, such as radio and television.
- The rise of the internet and social media platforms in the 21st century has further transformed the landscape of

parasocial interaction, blurring the lines between reality and fiction and fostering a sense of familiarity and closeness with media figures.

- Mass media has played a crucial role in shaping parasocial relationships throughout history, providing the means for audiences to connect with personalities and characters in a way that transcends physical proximity.
- Technological advancements and the emergence of social media have significantly impacted the way we experience and engage in parasocial interaction, with platforms such as Facebook, Twitter, Instagram, and YouTube providing unprecedented access to the lives of celebrities and influencers.
- Parasocial interaction has had a profound impact on society and culture, shaping the way we perceive and engage with public figures, influencing our own relationships and mental health, and giving rise to the phenomenon of celebrity worship.
- Potential negative consequences of parasocial interaction include a decline in face-to-face social interaction, unrealistic expectations and comparisons, and impacts on self-esteem and mental health.
- As we continue to navigate the digital age, it is essential for researchers, mental health professionals, and society as a whole to remain vigilant and adaptive in understanding the complexities of parasocial interaction, harnessing the positive aspects of these relationships while mitigating potential negative consequences.

2

PSYCHOLOGICAL FOUNDATIONS OF PARASOCIAL RELATIONSHIPS

An image depicting the emotional bond between a person and their favorite fictional character, showcasing the psychological foundation of their parasocial relationship.

Parasocial relationships refer to the one-sided emotional bonds that individuals form with media figures, such as celebrities, fictional characters, or social media influencers. While lacking direct interaction and reciprocity, these relationships can significantly impact an individual's emotions, attitudes, and behaviors. As our society becomes increasingly intertwined with various forms of media and technology, it is crucial to understand the psychological foundations that drive the formation and maintenance of parasocial relationships.

In this chapter, we will delve into the psychological underpinnings of parasocial relationships, exploring the role of attachment theory, cognitive processes, emotional responses, personality traits, and the impact of social media and technology on parasocial dynamics. By examining these factors, we aim to provide a comprehensive understanding of the complex interplay between an individual's psychological makeup and their propensity to form parasocial bonds.

To begin, we will discuss the role of attachment theory in parasocial relationships. Attachment theory, developed by John Bowlby and Mary Ainsworth, posits that humans have an innate need to form close emotional bonds with others, which serve as a secure base from which they can explore the world and seek comfort in times of distress. We will explore how this fundamental human need for attachment can manifest in the formation of parasocial relationships, particularly when an individual's attachment needs are not met through traditional interpersonal relationships.

Next, we will examine the cognitive processes and emotional responses involved in parasocial interactions. This section will delve into how individuals process information about media figures, the emotional responses elicited by parasocial interactions, and the factors that contribute to the intensity and longevity of these emotional bonds.

Following this, we will discuss the influence of personality traits on parasocial relationship formation. Research has shown that certain personality traits, such as extraversion, openness to experience, and neuroticism, can predispose individuals to form parasocial relationships. We will explore the mechanisms through which these traits influence parasocial dynamics and the implications of these findings for our understanding of individual differences in parasocial relationship formation.

In the subsequent section, we will investigate the impact of social media and technology on parasocial dynamics. The rise of social media platforms and advances in technology have significantly altered the landscape of parasocial interactions, allowing for more immersive and interactive experiences with media figures. We will discuss the ways in which these technological advancements have influenced the formation, maintenance, and dissolution of parasocial relationships.

Finally, we will conclude by synthesizing our understanding of the psychological foundations of parasocial relationships, highlighting the key takeaways from each section, and discussing the implications of this knowledge for future research and practical applications. By understanding the psychological underpinnings of parasocial relationships, we can better comprehend the complex and multifaceted nature of human connection in the modern media landscape.

The Role of Attachment Theory in Parasocial Relationships

Attachment theory, a cornerstone of modern psychology, provides a valuable framework for understanding the development and maintenance of parasocial relationships. Originally proposed by John Bowlby in the 1950s, attachment theory posits that humans have an innate need to form strong emotional bonds with others, primarily caregivers, to ensure survival and well-being. These early attachment experiences shape our expectations and behaviors in future relationships, both real and imagined.

In the context of parasocial relationships, attachment theory helps explain why individuals may form emotional connections with media figures, such as celebrities, fictional characters, or social media influencers. Just as we seek comfort, security, and validation from our primary caregivers, we may also derive these feelings from parasocial interactions. This section will delve into the ways attachment theory informs our understanding of parasocial relationships, including the role of secure and insecure attachment styles, the functions served by parasocial bonds, and the implications for mental health and well-being.

First, let's consider how attachment styles may influence parasocial relationship formation. Secure attachment, characterized by trust, emotional availability, and a sense of safety in relationships, may lead individuals to form healthy parasocial bonds that complement their real-life connections. For example, a securely attached person might admire a celebrity for their talent or achievements without relying on them for emotional support or self-worth.

In contrast, insecure attachment styles, such as anxious or avoidant, may predispose individuals to form more intense or dysfunctional parasocial relationships. An anxiously attached person might become overly preoccupied with a media figure, seeking constant reassurance and validation from their perceived connection. Meanwhile, an avoidantly attached individual may prefer parasocial relationships to real-life connections, as they offer a sense of control and emotional distance.

Beyond attachment styles, attachment theory also sheds light on the functions served by parasocial relationships. For some individuals, media figures may serve as attachment figures, providing a sense of security, comfort, and guidance. This can be particularly true for those who lack strong real-life attachments or face challenging life circumstances. For example, a lonely teenager might find solace in their favorite artist's music, feeling understood and supported by their lyrics. Similarly, a person struggling with mental health issues might find comfort in following a social media influencer who openly shares their own experiences and coping strategies.

While parasocial relationships can offer some benefits, it is essential to consider the potential drawbacks and implications for mental health. Overreliance on parasocial connections may hinder the development of real-life relationships, leading to social isolation and decreased well-being. Additionally, the one-sided nature of parasocial relationships can create unrealistic expectations and idealization of media figures, setting individuals up for disappointment and feelings of rejection.

In conclusion, attachment theory offers valuable insights into the psychological foundations of parasocial relationships. By examining the role of attachment styles and the functions served by parasocial bonds, we can better understand why individuals form these connections and the potential implications for their mental health and well-being. As our world becomes increasingly interconnected through social media and technology, it is crucial to continue exploring the complex interplay between attachment, parasocial relationships, and human connection.

Cognitive Processes and Emotional Responses in Parasocial Interactions

This section will delve into the cognitive processes and emotional responses that underlie parasocial interactions. Understanding these mechanisms is crucial to comprehending the formation and maintenance of parasocial relationships, as well as their potential impact on individuals' well-being and social functioning.

Cognitive processes refer to the mental activities that facilitate the acquisition, organization, and use of information. In the context of parasocial interactions, these processes involve the perception and interpretation of media characters' behaviors and the integration of this information into individuals' mental representations of these characters. Some of the key cognitive processes involved in parasocial interactions include attention, perception, memory, and mental simulation.

Attention plays a critical role in parasocial interactions, as individuals need to selectively focus on relevant aspects of media characters' behaviors and attributes. This selective attention can be influenced by factors such as individuals' interests, preferences, and emotional states.

For instance, a lonely person might be more likely to pay attention to a media character who displays warmth and friendliness.

Perception involves the interpretation of media characters' behaviors and attributes, which can be influenced by individuals' pre-existing beliefs, expectations, and stereotypes. For example, a person who holds a stereotype that all politicians are dishonest might be more likely to perceive a fictional politician's character as deceitful, even if their behavior is ambiguous.

Memory plays a crucial role in forming and maintaining parasocial relationships, as individuals need to store and retrieve information about media characters' behaviors, attributes, and interactions. This information can be organized in the form of mental schemas, which are cognitive structures that help individuals make sense of complex social information.

Mental simulation refers to the process of imagining hypothetical scenarios and events involving media characters. This cognitive process can facilitate emotional responses and feelings of connection towards these characters, as individuals can vicariously experience their emotions, thoughts, and actions.

Emotional responses in parasocial interactions can be diverse and complex, ranging from feelings of affection and admiration to anger and disappointment. These emotional responses can be influenced by factors such as individuals' attachment styles, personality traits, and social needs. For example, a person with a secure attachment style might be more likely to experience positive emotions towards a media character who displays trustworthiness and emotional availability.

Emotions can also serve as motivational forces that drive individuals to engage in parasocial interactions and maintain parasocial relationships. For instance, feelings of empathy and compassion can motivate individuals to seek out and follow the stories of media characters who face adversity or struggle with personal challenges. Similarly, feelings of admiration and identification can drive individuals to emulate the behaviors and attributes of media characters they perceive as role models.

In conclusion, cognitive processes and emotional responses play a crucial role in shaping individuals' parasocial interactions and relationships. By understanding these underlying mechanisms, we can gain valuable insights into the factors that contribute to the formation and maintenance of parasocial relationships, as well as their potential impact on individuals' well-being and social functioning.

The Influence of Personality Traits on Parasocial Relationship Formation

In this section, we delve into the fascinating world of personality traits and their impact on the formation of parasocial relationships. As we explore this topic, it is essential to remember that personality traits are complex and multifaceted, and their influence on parasocial interactions can vary significantly from one individual to another. Nonetheless, understanding the role of personality traits in parasocial relationship formation can provide valuable insights into why some people are more susceptible to developing these relationships than others.

To begin with, let us consider the Big Five personality traits, which are widely recognized as the primary dimensions of human personality. These traits include extraversion, agreeableness, conscientiousness, neuroticism, and openness to experience. Research has shown that certain personality traits can predispose individuals to form parasocial relationships more readily than others.

For instance, individuals with high levels of extraversion tend to be more outgoing, sociable, and assertive. They often seek social stimulation and enjoy engaging with others. In the context of parasocial relationships, extroverted individuals may be more likely to form connections with media figures as a means of satisfying their social needs. They may also be more prone to developing strong emotional attachments to these figures, as they are naturally inclined to seek out social connections.

On the other hand, individuals with high levels of neuroticism are characterized by emotional instability, anxiety, and moodiness. These individuals may be more susceptible to forming parasocial relationships as a coping mechanism for their emotional distress. By developing a connection with a media figure, they may find solace and comfort in the perceived stability and predictability of the relationship.

Another personality trait that can influence parasocial relationship formation is agreeableness. Highly agreeable individuals are typically warm, empathetic, and cooperative. They are more likely to form parasocial relationships due to their innate desire to connect with others and their ability to empathize with the emotions and experiences of media figures.

In contrast, individuals with low levels of conscientiousness may be less likely to form parasocial relationships. These individuals tend to be disorganized, impulsive, and less goal-oriented. As a result, they may be less inclined to invest time and emotional energy into developing a connection with a media figure.

Lastly, openness to experience can also play a role in parasocial relationship formation. Individuals with high levels of openness are characterized by their curiosity, creativity, and willingness to explore new ideas and experiences. These individuals may be more likely to form parasocial relationships due to their natural inclination to seek out novel and stimulating experiences, such as engaging with media figures who offer unique perspectives or challenge their existing beliefs.

In conclusion, personality traits can significantly influence the formation of parasocial relationships. By understanding the role of these traits, we can gain a deeper understanding of why certain individuals are more susceptible to developing these connections and how they may be affected by them. As we continue to explore the psychological foundations of parasocial relationships, it is crucial to recognize the complexity and individuality of the human personality and its impact on our interactions with the media figures we admire and follow.

The Impact of Social Media and Technology on Parasocial Dynamics

The advent of social media and advancements in technology have significantly transformed the landscape of parasocial relationships. In this section, we will delve into how these innovations have influenced the formation, maintenance, and dissolution of parasocial bonds, as well as the implications for individuals and society as a whole.

Social media platforms, such as Facebook, Instagram, Twitter, and YouTube, have provided unprecedented access to the lives of public figures, celebrities, and influencers. This increased accessibility has facilitated the development of parasocial relationships by allowing fans to feel more intimately connected to their favorite personalities. The interactive nature of social media enables users to engage with public figures through liking, commenting, and sharing their content, fostering a sense of reciprocity and closeness that was previously unattainable through traditional media channels.

Moreover, the rise of live streaming and video content has further blurred the lines between reality and fiction, as fans are now able to witness the daily lives of their favorite personalities in real time. This immersive experience can intensify the emotional investment in parasocial relationships, as fans may feel they are part of the public figure's life and privy to their most personal moments.

Technology has also played a crucial role in shaping parasocial dynamics. For instance, virtual reality (VR) and augmented reality (AR) have the potential to create even more immersive and interactive experiences, allowing users to feel as though they are physically present with their favorite personalities. This heightened sense of presence can further strengthen parasocial bonds and blur the distinction between real and parasocial relationships.

While the impact of social media and technology on parasocial dynamics has undoubtedly facilitated the formation of these relationships, it is essential to consider the potential consequences. The constant exposure to the lives of public figures can lead to social comparison, which may result in feelings of inadequacy, envy, and even depression. Additionally, the time and emotional investment in parasocial relationships may detract from developing and maintaining real-life social connections.

Furthermore, the dissolution of parasocial relationships can be particularly distressing in the digital age. Public figures may choose to withdraw from social media or alter their online persona, leaving fans feeling abandoned or betrayed. In extreme cases, the loss of a parasocial relationship can lead to grief-like symptoms as fans mourn the loss of their imagined connection.

In conclusion, the impact of social media and technology on parasocial dynamics is multifaceted and complex. While these innovations have undoubtedly facilitated the formation and maintenance of parasocial relationships, it is crucial to consider the potential consequences for individuals and society. As we continue to navigate the ever-evolving digital landscape, we must foster a healthy balance between our online and offline lives and prioritize genuine social connections over parasocial bonds.

Understanding the Psychological Underpinnings of Parasocial Relationships

In conclusion, the psychological foundations of parasocial relationships are complex and multifaceted, encompassing various aspects of human cognition, emotion, and behavior. By examining these underpinnings, we can better understand the nature of parasocial interactions and how they impact our lives in the digital age.

Attachment theory has provided valuable insights into the formation and maintenance of parasocial relationships, highlighting the role of secure and insecure attachment styles in shaping individuals' tendencies to engage in parasocial interactions. This understanding can help us recognize the potential benefits and pitfalls of parasocial relationships and inform interventions aimed at promoting healthy attachment patterns and reducing reliance on parasocial connections.

Cognitive processes and emotional responses play a crucial role in developing and sustaining parasocial relationships. The human tendency to anthropomorphize and engage in social projection allows us to form emotional bonds with media figures, while the emotional rewards we derive from these relationships can further reinforce our attachment to them. Recognizing these cognitive and emotional dynamics can help us navigate the complex landscape of parasocial interactions and make more informed choices about our media consumption.

Personality traits also significantly influence the formation and intensity of parasocial relationships. Individuals with certain personality characteristics, such as high levels of empathy, openness to experience, and extraversion, may be more prone to forming parasocial attachments. Understanding the role of personality in parasocial dynamics can help us identify individuals who may be more susceptible to the potential negative effects of parasocial relationships and develop targeted interventions to promote healthy media habits.

The impact of social media and technology on parasocial dynamics cannot be overstated. The ubiquity of social media platforms and the increasing interactivity of media content have blurred the lines between real and parasocial relationships, making it easier than ever to form and maintain parasocial connections. As we continue to navigate this rapidly evolving digital landscape, it is essential to be aware of the psychological underpinnings of parasocial relationships and their potential implications for our well-being.

In summary, understanding the psychological foundations of parasocial relationships is crucial for both individuals and society as a whole. By examining the various factors that contribute to forming and maintaining these relationships, we can make more informed decisions about our media consumption, foster healthier connections with media figures and enhance our overall well-being. As we continue to explore the complexities of parasocial interactions, we can develop a more nuanced understanding of the human experience in the digital age and harness the power of media to enrich our lives.

Chapter Summary

- Parasocial relationships refer to the one-sided emotional bonds that individuals form with media figures, such as celebrities, fictional characters, or social media influencers.
- Attachment theory, developed by John Bowlby and Mary Ainsworth, posits that humans have an innate need to form close emotional bonds with others. This theory helps explain why individuals may form emotional connections with media figures, particularly when their attachment needs are not met through traditional interpersonal relationships.
- Cognitive processes, such as attention, perception, memory, and mental simulation, play a crucial role in parasocial interactions, as they facilitate the acquisition, organization, and use of information about media figures.
- Emotional responses in parasocial interactions can be diverse and complex, ranging from feelings of affection and admiration to anger and disappointment. These emotional responses can be influenced by factors such as individuals' attachment styles, personality traits, and social needs.
- Certain personality traits, such as extraversion, openness to experience, and neuroticism, can predispose individuals to form parasocial relationships. Understanding the role of these traits can help identify individuals who may be more susceptible to the potential negative effects of parasocial relationships.
- The rise of social media platforms and advances in technology have significantly altered the landscape of parasocial interactions, allowing for more immersive and interactive experiences with media figures.
- The constant exposure to the lives of public figures through social media can lead to social comparison, which may result in feelings of inadequacy, envy, and even depression.

Additionally, the time and emotional investment in parasocial relationships may detract from the development and maintenance of real-life social connections.

- Understanding the psychological foundations of parasocial relationships is crucial for both individuals and society as a whole, as it can help inform interventions aimed at promoting healthy attachment patterns, reducing reliance on parasocial connections, and fostering healthier connections with media figures.

3

THE ROLE OF MEDIA AND TECHNOLOGY IN PARASOCIAL INTERACTION

An image depicting a person engaging in a deep conversation with their favorite virtual influencer, showcasing the blurred lines between reality and the digital world in parasocial interactions.

Parasocial interaction refers to the one-sided relationships that individuals form with media figures, such as celebrities, television characters, and social media influencers. These relationships, while not reciprocal in nature, can evoke strong emotional responses and attachments in the audience. Over the years, the evolution of media and technology has significantly shaped the nature and intensity of parasocial interactions.

In the early days of mass communication, traditional media platforms such as television, radio, and print were the primary means through which audiences engaged with media figures. These platforms allowed for limited interaction, with audiences passively consuming content and developing attachments to the characters and personalities they encountered. The parasocial relationships formed during this era were largely based on the curated personas and narratives presented by the media.

The advent of the internet and the rise of social media have dramatically transformed the landscape of parasocial interaction. Platforms such as Facebook, Twitter, Instagram, and YouTube have enabled media figures to engage with their audiences more directly and personally. This increased accessibility and interactivity have blurred the lines between public and private spheres, fostering a sense of intimacy and familiarity between media figures and their followers. As a result, the parasocial relationships formed in the digital age are often more intense and complex than those of the past.

Looking ahead, emerging technologies such as virtual reality and artificial intelligence have the potential to further revolutionize the nature of parasocial interaction. These technologies can create immersive and interactive experiences that may deepen the emotional connections between individuals and media figures, raising new questions about the psychological and ethical implications of technology-driven parasocial relationships.

In this chapter, we will explore the various ways in which media and technology have shaped and continue to influence the phenomenon of parasocial interaction. We will begin by examining the role of traditional media platforms in fostering parasocial relationships before delving into the impact of social media on the formation and maintenance of these connections. Next, we will discuss the potential implications of virtual reality and artificial intelligence on the future of parasocial interaction and the ethical considerations that arise from these advancements. Finally, we will conclude by reflecting on the benefits and challenges of media and technology in the realm of parasocial interaction and consider how we might navigate this ever-evolving landscape in a responsible and conscientious manner.

Traditional Media Platforms: Television, Radio, and Print

This section will delve into the role of traditional media platforms such as television, radio, and print in shaping parasocial interaction. These platforms have been the primary means of communication and entertainment for decades and have significantly influenced the way people form connections with public figures and fictional characters.

Television, as a visual medium, has played a pivotal role in fostering parasocial relationships. Since its inception, TV shows and movies have captivated audiences, creating a sense of familiarity and attachment to the characters portrayed on screen. This emotional connection is further strengthened by the serialized nature of many television programs, which allows viewers to follow the lives of their favorite characters over an extended period. Consequently, viewers often develop strong emotional bonds with these characters, experiencing feelings of friendship, love, or even enmity, despite the one-sided nature of the relationship.

Radio, on the other hand, relies on auditory communication to engage its audience. While it may lack the visual component of television, radio has its unique charm in fostering parasocial interaction. The power of the human voice, combined with the listener's imagination, can create a strong sense of intimacy and connection. Radio personalities, through their distinctive voices and conversational styles, can make listeners feel as though they are part of an ongoing dialogue, even though the interaction is entirely one-sided. This sense of connection is further amplified by the live nature of many radio broadcasts, which allows for a semblance of real-time interaction between the host and the audience.

Print media, which includes newspapers, magazines, and books, has also played a significant role in shaping parasocial interaction. While it may not offer the same immediacy and intimacy as television or radio, print media allows readers to connect with authors, journalists, and fictional characters through the written word. The immersive nature of reading enables individuals to become emotionally invested in the stories and lives of the characters they encounter, often leading to a strong sense of attachment and identification.

Despite the differences in their modes of communication, all three traditional media platforms share a common characteristic: they are primarily unidirectional, meaning that the audience receives information without the ability to interact directly with the content creators or characters. This unidirectional nature has been a defining feature of parasocial interaction, as it allows individuals to form emotional connections without the need for reciprocal communication.

In conclusion, traditional media platforms such as television, radio, and print have played a crucial role in shaping parasocial interaction. They have provided the means for audiences to form emotional bonds with public figures and fictional characters, despite the one-sided nature of the relationship. As we move forward, it is essential to recognize the impact of these traditional platforms on parasocial interaction as they continue to influence the way we connect with others in an increasingly digital world.

The Rise of Social Media: A New Frontier for Parasocial Relationships

The advent of social media has revolutionized the way we communicate, connect, and interact with one another. It has also given rise to a new frontier for parasocial relationships, providing an unprecedented level of access to the lives of celebrities, influencers, and public figures. In this section, we will explore the role of social media in shaping parasocial interactions and the unique characteristics that set it apart from traditional media platforms.

One of the most significant aspects of social media is its ability to blur the lines between public and private spheres. Unlike traditional media, where celebrities and public figures maintain a certain level of distance from their audience, social media platforms such as Instagram, Twitter, and Facebook allow users to follow, like, and comment on the lives of their favorite personalities in real time. This increased level of intimacy and accessibility fosters a sense of familiarity and closeness, which can lead to the development of parasocial relationships.

Moreover, social media platforms enable users to engage in two-way communication with celebrities and influencers. This interactive nature of social media creates an illusion of reciprocity, as fans may feel that they are part of a genuine conversation or relationship with their favorite personalities. In reality, however, these interactions are often one-sided and carefully curated by the celebrity or influencer to maintain their public image and brand.

Another key factor that contributes to the rise of parasocial relationships on social media is the concept of *"FOMO"* or the fear of missing out. Social media platforms are designed to keep users constantly updated on the latest news, trends, and events, which can lead to a sense of urgency and a need to stay connected at all times. This constant exposure to the lives of celebrities and influencers can further fuel the development of parasocial relationships, as users may feel compelled to keep up with their favorite personalities and engage in their online activities.

The rise of social media influencers has also played a significant role in shaping parasocial interactions. Influencers, who are often regular individuals who have gained a large following due to their expertise or relatability, can create a sense of authenticity and trust with their audience. This perceived authenticity can lead to stronger parasocial relationships, as users may feel a deeper connection with influencers compared to traditional celebrities who may seem more distant and unattainable.

While social media has undoubtedly opened up new avenues for parasocial relationships, it is essential to recognize the potential risks and challenges associated with this phenomenon. The constant exposure to the lives of others can lead to feelings of inadequacy, envy, and even addiction, as users may become overly invested in their online relationships at the expense of their real-life connections. Additionally, the curated nature of social media can create unrealistic expectations and standards, further exacerbating these negative emotions.

In conclusion, the rise of social media has significantly impacted the landscape of parasocial interaction, providing new opportunities for connection and engagement with celebrities, influencers, and public figures. As we continue to navigate this new frontier, it is crucial to strike a balance between the benefits and challenges of social media-driven parasocial relationships, ensuring that we maintain a healthy and realistic perspective on our online connections.

Virtual Reality and Artificial Intelligence: The Future of Parasocial Interaction

As we venture further into the 21st century, the rapid advancements in virtual reality (VR) and artificial intelligence (AI) are poised to reshape the landscape of parasocial interaction. These cutting-edge technologies have the potential to create immersive and highly personalized experiences, blurring the lines between reality and the virtual world. This section will explore how VR and AI are revolutionizing parasocial interaction and the implications of these developments for the future.

Virtual reality, a technology that allows users to immerse themselves in a computer-generated environment, has been gaining traction in various industries, including entertainment, education, and healthcare. In parasocial interaction, VR offers a unique opportunity for individuals to engage with their favorite celebrities, fictional characters, or influencers more intimately and interactively. For instance, fans can attend virtual concerts, participate in meet-and-greets, or even explore the personal lives of their idols through VR experiences. This heightened level of engagement can lead to stronger parasocial bonds and a greater sense of connection with the virtual persona.

On the other hand, artificial intelligence has made significant strides in recent years, particularly in the field of natural language processing and machine learning. This has led to the development of sophisticated chatbots and virtual assistants that can simulate human-like conversations and interactions. As AI continues to improve, it is not far-fetched to imagine a future where individuals form parasocial relationships with AI-powered virtual beings. These AI personas can be tailored to the user's preferences, providing a highly personalized and adaptive parasocial experience. In fact, some users may even prefer the company of these virtual companions over real-life social interactions, given their ability to cater to the user's needs and desires without judgment or conflict.

While the potential benefits of VR and AI in enhancing parasocial interaction are undeniable, it is crucial to consider the potential drawbacks and ethical concerns that may arise. For instance, the immersive nature of VR can lead to addiction and social isolation, as users may prioritize their virtual relationships over real-life connections. Additionally, the highly personalized nature of AI-driven parasocial interaction raises concerns about privacy and data security, as users may inadvertently share sensitive information with their virtual companions.

Moreover, the increasing sophistication of AI and VR technologies may blur the line between reality and the virtual world, making it difficult for users to distinguish between genuine human connections and parasocial relationships. This could lead to an overreliance on virtual interactions, potentially hindering the development of essential social skills and emotional intelligence.

In conclusion, virtual reality and artificial intelligence hold great promise in revolutionizing the future of parasocial interaction. However, it is essential to strike a balance between leveraging the benefits of these technologies and addressing the potential challenges they pose. As we continue to explore the possibilities of VR and AI in shaping parasocial relationships, it is crucial to remain mindful of the ethical considerations and implications of these advancements, ensuring that we harness their potential responsibly and sustainably.

Ethical Considerations and Implications of Technology-Driven Parasocial Interaction

As we delve deeper into technology-driven parasocial interaction, it is crucial to address the ethical considerations and implications that arise from these advancements. While the benefits of parasocial relationships are evident, such as providing companionship and entertainment, there are also potential negative consequences that warrant our attention. In this section, we will explore the ethical concerns surrounding the use of media and technology in fostering parasocial interactions and discuss potential solutions to mitigate these issues.

One of the primary ethical concerns in technology-driven parasocial interaction is the potential for exploitation and manipulation. With the increasing sophistication of artificial intelligence and virtual reality, it is becoming easier for creators to design highly realistic and immersive experiences that tap into our innate desire for connection. This can lead to users becoming overly attached to virtual characters or influencers, making them vulnerable to manipulation for financial gain or other nefarious purposes. To address this issue, it is essential for regulators and industry professionals to establish guidelines and ethical standards that prioritize user well-being and prevent exploitation.

Another ethical consideration is the impact of parasocial relationships on mental health. While these interactions can provide comfort and companionship, they may also contribute to feelings of isolation and loneliness, as users may prioritize virtual relationships over real-life connections. Additionally, the curated nature of online personas can lead to unrealistic expectations and social comparison, potentially exacerbating mental health issues such as anxiety and depression. To mitigate these risks, users need to maintain a balanced approach to their engagement with virtual characters and influencers, ensuring that they maintain healthy real-life relationships and seek professional help if necessary.

Privacy and data security are also significant ethical concerns in technology-driven parasocial interaction. As users engage with virtual characters and influencers, they may inadvertently share personal information that malicious actors can exploit. This raises questions about the responsibility of creators and platforms to protect user data and ensure that their virtual experiences are secure. To address this issue, industry professionals must prioritize data security and privacy in developing new technologies and platforms while educating users about the potential risks associated with sharing personal information online.

Lastly, the issue of representation and diversity in virtual characters and influencers is an important ethical consideration. As parasocial relationships become increasingly prevalent, it is essential that the virtual world reflects the diversity of our real world, ensuring that all users feel represented and included. This involves not only the creation of diverse characters but also the consideration of cultural sensitivity and the avoidance of harmful stereotypes. By fostering a more inclusive virtual environment, we can promote empathy and understanding among users, ultimately contributing to a more harmonious society.

In conclusion, the ethical considerations and implications of technology-driven parasocial interaction are complex and multifaceted. By acknowledging these concerns and proactively addressing them, we can ensure that the benefits of parasocial relationships are maximized while minimizing potential harm. This requires collaboration between industry professionals, regulators, and users, as well as a commitment to prioritizing the well-being of individuals and society as a whole.

Balancing the Benefits and Challenges of Media and Technology in Parasocial Interaction

In conclusion, the role of media and technology in parasocial interaction has evolved significantly over the years, transforming the way individuals form and maintain relationships with public figures and fictional characters. From traditional media platforms such as television, radio, and print to the rise of social media, virtual reality, and artificial intelligence, the landscape of parasocial interaction has expanded and diversified, offering new opportunities for connection and engagement.

On the one hand, these advancements in media and technology have provided numerous benefits to both audiences and public figures. For audiences, parasocial interactions can offer a sense of companionship, emotional support, and even educational value. They can help individuals feel more connected to the world around them and foster a sense of belonging within various communities. For public figures, parasocial relationships can serve as a powerful tool for building and maintaining a fan base, promoting their work, and engaging with their audience in a more personal and authentic manner.

However, the increasing prevalence of technology-driven parasocial interaction also presents several challenges and ethical considerations. The potential for addiction, negative mental health effects, and the blurring of boundaries between reality and fiction are all concerns that must be addressed as we continue to navigate this evolving landscape. Additionally, the potential for exploitation and manipulation of audiences by public figures, as well as the ethical implications of artificial intelligence and virtual reality in fostering parasocial relationships, must be carefully considered.

Ultimately, the key to balancing the benefits and challenges of media and technology in parasocial interaction lies in promoting responsible consumption and engagement. This includes fostering media literacy and critical thinking skills among audiences, encouraging public figures to maintain ethical standards in their interactions with fans, and supporting ongoing research and dialogue on the impact of technology on parasocial relationships. By doing so, we can ensure that the positive aspects of parasocial interaction continue to thrive while mitigating the potential risks and negative consequences.

As an author and industry expert, I hope this chapter has provided valuable insights into the complex and ever-evolving world of parasocial interaction. By understanding the role of media and technology in shaping these relationships, we can better appreciate the nuances of our connections with public figures and fictional characters and navigate the challenges and opportunities that lie ahead.

Chapter Summary

- Parasocial interaction refers to the one-sided relationships that individuals form with media figures, such as celebrities, television characters, and social media influencers.
- Traditional media platforms like television, radio, and print have played a significant role in shaping parasocial interactions, allowing audiences to develop attachments to the characters and personalities they encounter.
- The advent of the internet and social media has transformed the landscape of parasocial interaction, enabling media figures to engage with their audiences more directly and personally, fostering a sense of intimacy and familiarity.
- Virtual reality and artificial intelligence have the potential to revolutionize parasocial interaction by creating immersive and interactive experiences that may deepen emotional connections between individuals and media figures.
- Ethical considerations in technology-driven parasocial interaction include potential exploitation and manipulation, negative mental health effects, privacy and data security concerns, and the need for representation and diversity in virtual characters and influencers.
- Balancing the benefits and challenges of media and technology in parasocial interaction requires promoting responsible consumption and engagement, fostering media literacy and critical thinking skills, and encouraging public figures to maintain ethical standards in their interactions with fans.
- Addressing the ethical concerns surrounding technology-driven parasocial interaction involves collaboration between industry professionals, regulators, and users, as well as a

commitment to prioritizing the well-being of individuals and society as a whole.

- As media and technology continue to evolve, it is crucial to understand their role in shaping parasocial relationships and navigate the challenges and opportunities that lie ahead in a responsible and conscientious manner.

4

CELEBRITIES, INFLUENCERS, AND PARASOCIAL BONDS

An image of a futuristic red carpet event where celebrities, influencers, and their virtual avatars are mingling together, showcasing the evolution of parasocial bonds in the digital age.

In today's hyper-connected world, the lines between reality and virtuality have become increasingly blurred. As we spend more time online, we find ourselves forming connections with people we have never met in person. These connections, known as parasocial relationships, have become a pervasive aspect of modern life, particularly in the realm of celebrities and influencers. In this chapter, we will explore the rise of parasocial relationships in the digital age, delving into the allure of celebrities, the power of influencers, and the psychology behind these unique bonds.

Parasocial relationships are one-sided connections in which an individual feels a sense of attachment to a media figure, such as a celebrity or influencer, without any direct interaction or reciprocation from the figure. These relationships have existed for decades, with fans forming attachments to movie stars, musicians, and other public figures. However, the advent of social media and the internet has amplified the prevalence and intensity of these relationships, making it easier than ever for individuals to feel connected to their favorite celebrities and influencers.

The digital age has provided a platform for celebrities and influencers to share their lives with the world, offering fans unprecedented access and intimacy. Through social media platforms like Instagram, Twitter, and YouTube, fans can follow their favorite stars' daily activities, engage with their content, and even communicate with them directly. This increased accessibility has fostered a sense of familiarity and closeness between fans and celebrities, leading to stronger parasocial bonds.

As we navigate the complexities of these relationships, it is essential to understand the factors that contribute to their formation and the impact they have on our lives. In the following sections, we will examine the allure of celebrities and the role of fame and status in shaping parasocial bonds. We will also discuss the power of influencers and the importance of authenticity in building connections with their audience.

Furthermore, we will consider the future of celebrities, influencers, and parasocial bonds in a connected world. As technology continues to evolve and our lives become increasingly intertwined with the digital realm, it is crucial to understand the implications of these relationships on our well-being and the ways in which we can adapt to maintain a healthy balance between our online and offline lives.

The Allure of Celebrities: How Fame and Status Shape Parasocial Bonds

In today's media-saturated world, celebrities have become an integral part of our cultural landscape. From movie stars to pop icons, these individuals often hold a unique and powerful position in society, capturing the attention and admiration of millions. But what is it about celebrities that makes them so alluring to the masses? And how do their fame and status shape the parasocial bonds that we form with them? In this section, we will delve into the fascinating world of celebrity culture and explore the psychological underpinnings of our attraction to these larger-than-life figures.

One of the key factors that contribute to the allure of celebrities is the concept of social comparison. As humans, we have an innate tendency to compare ourselves to others, particularly those who we perceive as being more successful, attractive, or talented than ourselves. By virtue of their fame and status, celebrities often embody these aspirational qualities, making them prime targets for social comparison. This can lead to the formation of parasocial bonds as we seek to emulate or identify with these individuals in an attempt to enhance our own self-esteem and sense of self-worth.

Another important aspect of celebrity allure is the role of fantasy and escapism. Celebrities often lead lives that are vastly different from our own, filled with glamour, excitement, and seemingly endless opportunities. This can make them an attractive source of vicarious pleasure as we immerse ourselves in their world and temporarily escape the mundane realities of our own lives. Through parasocial bonds, we can experience a sense of connection and intimacy with these individuals, allowing us to share in their triumphs and tribulations, even if only in our own minds.

The media also plays a significant role in shaping our attraction to celebrities and the formation of parasocial bonds. Through constant exposure to images, stories, and interviews, we are given a carefully curated glimpse into the lives of these individuals, often presenting them as relatable, down-to-earth, and accessible. This can foster a sense of familiarity and closeness, making it easier for us to form parasocial attachments to these figures. Additionally, the media often perpetuates narratives of celebrity success and transformation, reinforcing the idea that fame and status are attainable goals for those who are willing to work hard and persevere.

The formation of parasocial bonds with celebrities can also be influenced by our own individual needs and desires. For some, these relationships may provide a sense of belonging and validation, particularly if they feel isolated or marginalized in their own lives. For others, celebrities may serve as role models or sources of inspiration, motivating them to pursue their own dreams and aspirations. In either case, the parasocial bond can serve as a powerful psychological resource, helping to fulfill emotional and cognitive needs that may not be met through traditional social relationships.

In conclusion, the allure of celebrities and the parasocial bonds that we form with them can be attributed to a complex interplay of psychological, social, and cultural factors. From social comparison and escapism to media influence and personal needs, our attraction to these individuals is both multifaceted and deeply ingrained in our human nature. As we continue to navigate the ever-evolving landscape of celebrity culture, it is important to recognize and understand the underlying mechanisms that drive our fascination with fame and status, as well as the potential benefits and drawbacks of these parasocial relationships.

Influencers and the Power of Social Media: Building Connections through Authenticity

In today's digital landscape, social media platforms have become the primary stage for the emergence of a new breed of celebrities - influencers. These individuals, often ordinary people who have amassed a significant following on platforms like Instagram, YouTube, and TikTok, have managed to create a unique space for themselves in the world of parasocial interactions. Unlike traditional celebrities, influencers often rely on authenticity and relatability to forge strong connections with their audience, making their parasocial bonds even more potent.

The power of social media lies in its ability to provide a direct line of communication between influencers and their followers. This accessibility allows influencers to share their lives, thoughts, and experiences with their audience in real time, creating a sense of intimacy that is hard to replicate in traditional media. By showcasing their genuine selves, influencers can foster a sense of trust and loyalty among their followers, who often view them as friends or confidants rather than distant, unattainable figures.

One of the key factors that contribute to the success of influencers in building parasocial relationships is their ability to create content that resonates with their audience on a personal level. By sharing their own experiences, struggles, and triumphs, influencers can make their followers feel seen and understood, further strengthening the emotional bond between them. This relatability is particularly important for younger audiences, who often turn to social media as a source of guidance and support in navigating the challenges of adolescence and young adulthood.

Another crucial aspect of the influencer-follower dynamic is the sense of community that can develop around a particular influencer. Through their content, influencers can create a shared identity and values that their followers can rally around, fostering a sense of belonging and camaraderie. This community-building aspect of parasocial relationships can have significant positive effects on individuals' well-being, as it provides them with a supportive network of like-minded individuals who share their interests and passions.

However, the authenticity that is so central to the influencer-follower relationship can also be a double-edged sword. As influencers strive to maintain their image of relatability and genuineness, they may feel pressured to constantly share their lives and experiences with their audience, blurring the lines between their public and private selves. This can lead to burnout, mental health struggles, and even the commodification of personal experiences for the sake of content creation.

In conclusion, the rise of influencers and the power of social media have significantly transformed the landscape of parasocial interactions, with authenticity and relatability becoming key drivers of these relationships. As we continue to navigate the complexities of our increasingly connected world, we must recognize the potential benefits and drawbacks of these parasocial bonds and foster a healthy and balanced approach to our engagement with celebrities, influencers, and the digital spaces they inhabit.

Navigating the Complexities of Parasocial Relationships: Benefits and Drawbacks

As we delve deeper into parasocial relationships, it becomes increasingly important to understand the complexities that come with these bonds. While there are undoubtedly numerous benefits to forming connections with celebrities and influencers, there are also potential drawbacks that must be considered. In this section, we will explore both the positive and negative aspects of parasocial relationships and discuss how individuals can navigate these complexities in a healthy and balanced manner.

On the one hand, parasocial relationships can offer a sense of companionship and support for individuals who may feel isolated or disconnected from their immediate social circles. On the other hand, celebrities and influencers often share personal stories, experiences, and emotions with their followers, fostering a sense of intimacy and understanding. This can lead to a feeling of camaraderie and connection, as fans may see themselves reflected in the lives of their favorite public figures.

Moreover, parasocial bonds can inspire and motivate personal growth and self-improvement. By observing the successes, challenges, and resilience of celebrities and influencers, fans may feel encouraged to pursue their own goals and aspirations. Additionally, these public figures often use their platforms to promote positive messages, such as self-love, mental health awareness, and social justice, which can inspire and empower their followers.

However, there are also potential drawbacks to consider when engaging in parasocial relationships. One of the primary concerns is the potential for over-identification with a celebrity or influencer, which can lead to an unhealthy level of emotional investment. This may result in feelings of disappointment or betrayal when the public figure does not meet the expectations of their followers or when they make choices that are not aligned with the values of their fans.

Another potential pitfall of parasocial relationships is the risk of neglecting one's real-life relationships and responsibilities. As individuals become increasingly absorbed in the lives of their favorite celebrities and influencers, they may inadvertently prioritize these parasocial bonds over their actual connections with friends, family, and loved ones. This can lead to isolation, loneliness, and disconnection from one's own life and experiences.

To navigate the complexities of parasocial relationships, individuals need to maintain a sense of balance and perspective. While it is natural to feel a connection with public figures who share their lives and experiences, it is important to remember that these relationships are ultimately one-sided and should not replace genuine, reciprocal connections with others. By engaging in a variety of social activities, maintaining open communication with loved ones, and setting boundaries around one's consumption of celebrity and influencer content, individuals can enjoy the benefits of parasocial bonds without succumbing to their potential drawbacks.

In conclusion, the world of celebrities, influencers, and parasocial relationships is a complex and ever-evolving landscape. As we continue to forge connections with public figures in the digital age, it is crucial to remain mindful of the benefits and drawbacks of these bonds and to navigate them with intention and self-awareness. By doing so, we can foster a healthy and balanced approach to our engagement with the celebrities and influencers who inspire and entertain us.

The Future of Celebrities, Influencers, and Parasocial Bonds in a Connected World

As we continue to navigate the ever-evolving digital landscape, the role of celebrities, influencers, and parasocial bonds in our lives will undoubtedly change and adapt. The rapid advancements in technology and the increasing interconnectedness of our world have created a unique environment where parasocial relationships can thrive. In this conclusion, we will explore the potential future of these relationships and consider the implications for both individuals and society as a whole.

One significant development that will likely shape the future of parasocial bonds is the increasing sophistication of artificial intelligence (AI) and virtual reality (VR) technologies. As these technologies become more advanced and accessible, it is possible that we will see a rise in virtual celebrities and influencers who exist solely in the digital realm. These virtual personas may offer even more opportunities for parasocial interaction, as they can be specifically designed to cater to the desires and preferences of their audience. This could lead to even stronger parasocial bonds, as individuals may feel a heightened sense of connection and intimacy with these virtual entities.

However, this increased reliance on virtual relationships also raises important ethical and psychological questions. As individuals become more emotionally invested in their parasocial connections, there is a risk that they may neglect or undervalue their real-life relationships. This could lead to increased feelings of isolation, loneliness, and a diminished sense of self-worth. It is crucial for individuals to strike a healthy balance between their online and offline lives and to recognize the limitations of parasocial bonds.

Another potential development in the future of parasocial relationships is the increasing awareness of the power dynamics at play in these connections. As more people become educated about the psychological mechanisms underlying parasocial bonds, they may become more critical of the ways in which celebrities and influencers use their platforms to manipulate and exploit their audiences. This increased scrutiny could lead to a demand for greater transparency and authenticity from public figures, as well as a shift in the way that fame and influence are perceived and valued in our society.

Finally, as our understanding of parasocial relationships continues to grow, there may be opportunities to harness the positive aspects of these bonds for the greater good. For example, celebrities and influencers could use their platforms to promote social causes, raise awareness about important issues, or encourage their followers to engage in prosocial behaviors. By leveraging the power of parasocial bonds in this way, public figures have the potential to create meaningful change and contribute to the betterment of society.

In conclusion, the future of celebrities, influencers, and parasocial bonds in a connected world is uncertain, but it is clear that these relationships will continue to play a significant role in our lives. As we move forward, we must remain mindful of the potential benefits and drawbacks of parasocial connections and strive to create a healthy balance between our online and offline relationships. By doing so, we can ensure that the digital age continues to enrich our lives rather than detract from them.

Chapter Summary

- Parasocial relationships are one-sided connections with media figures, such as celebrities and influencers, that have become more prevalent and intense in the digital age due to increased accessibility and intimacy through social media platforms.
- The allure of celebrities can be attributed to social comparison, fantasy and escapism, media influence, and personal needs and desires. These factors contribute to the formation of parasocial bonds and our fascination with fame and status.
- Influencers, a new breed of celebrities, rely on authenticity and relatability to forge strong connections with their audience. Social media platforms provide a direct line of communication, allowing influencers to share their lives and experiences, fostering trust and loyalty among their followers.
- Parasocial relationships can offer benefits such as companionship, support, and motivation. However, potential drawbacks include over-identification, unhealthy emotional investment, and neglect of real-life relationships and responsibilities.
- To navigate the complexities of parasocial relationships, individuals should maintain a sense of balance and perspective, engage in a variety of social activities, maintain open communication with loved ones, and set boundaries around their consumption of celebrity and influencer content.
- The future of parasocial bonds may be shaped by advancements in artificial intelligence and virtual reality technologies, leading to the rise of virtual celebrities and influencers. This raises ethical and psychological questions

about the impact of virtual relationships on individuals and society.

- As awareness of the power dynamics in parasocial relationships increases, there may be a demand for greater transparency and authenticity from public figures. Harnessing the positive aspects of these bonds for social causes and prosocial behaviors could contribute to meaningful change and the betterment of society.

5

THE IMPACT OF PARASOCIAL INTERACTION ON MENTAL HEALTH

An image depicting a person sitting alone in their room, surrounded by screens displaying their favorite social media influencers, while a shadow representing their mental health struggles looms in the background.

In today's digital age, the way we connect with others has evolved significantly. With the rise of social media platforms and the internet, we are now able to form connections with people we may never meet in person. One such phenomenon that has gained prominence in recent years is parasocial interaction (PSI). In this chapter, we will delve into the world of parasocial interaction, exploring its impact on mental health and providing insights on how to navigate this complex landscape.

Parasocial interaction refers to the one-sided relationships that individuals form with media figures, such as celebrities, influencers, or fictional characters. These relationships often mimic the emotional bonds and connections we form with people in our real lives. However, unlike traditional relationships, parasocial interactions are not reciprocal, meaning the media figure is typically unaware of the individual's existence.

The concept of parasocial interaction was first introduced by Horton and Wohl in 1956, as they observed the relationships that television viewers formed with characters and personalities on screen. Since then, the phenomenon has grown exponentially with the advent of social media, allowing individuals to feel even more connected to their favorite media figures.

The relevance of parasocial interaction to mental health is a topic that has garnered significant attention in recent years. As we spend more time online and engage with various media figures, it is crucial to understand the potential effects of these interactions on our mental well-being.

In this chapter, we will explore the positive and negative consequences of parasocial interaction on mental health. We will discuss how these interactions can provide a sense of comfort, companionship, and even inspiration, while also examining the potential pitfalls, such as feelings of inadequacy, loneliness, and obsession.

Furthermore, we will delve into strategies for balancing parasocial interaction to achieve optimal mental health outcomes. This includes recognizing the signs of unhealthy parasocial relationships, setting boundaries, and fostering genuine connections with others in our lives.

Finally, we will look toward the future of parasocial interaction and mental health, considering the ever-evolving landscape of technology and media consumption. As we continue to navigate this digital age, we must remain informed and adaptable, ensuring that we maintain a healthy balance between our online and offline lives.

Join us on this journey as we explore the fascinating world of parasocial interaction and its impact on mental health, providing valuable insights and guidance for navigating this complex and ever-evolving landscape.

The Positive Effects of Parasocial Interaction on Mental Health

Parasocial interaction, the one-sided relationship that individuals form with media figures, has become an increasingly prevalent aspect of modern life. With the rise of social media and the growing accessibility of celebrities and influencers, these relationships have the potential to significantly impact our mental health. While parasocial interaction is often viewed with skepticism, it is important to recognize that it can also positively affect mental health. In this section, we will explore the various ways in which parasocial interaction can contribute to improved mental health outcomes.

One of the most significant benefits of parasocial interaction is the sense of social support and companionship it can provide. For individuals experiencing loneliness or social isolation, engaging with a media figure can offer a sense of connection and belonging. This can be particularly beneficial for those who struggle to form connections in their offline lives, such as individuals with social anxiety or other mental health challenges. By providing a sense of social support, parasocial interaction can help to alleviate feelings of loneliness and improve overall mental well-being.

Another positive aspect of parasocial interaction is the opportunity to identify with and learn from media figures who serve as role models. By observing the actions and behaviors of these individuals, we can gain inspiration and motivation to pursue our own goals and aspirations. This can be particularly beneficial for individuals who may lack positive role models in their offline lives, as it offers an alternative source of guidance and support. Additionally, the achievements and successes of media figures can provide a sense of hope and optimism, which can contribute to improved mental health outcomes.

Parasocial interaction can also be a valuable tool for emotional regulation and coping. Engaging with media figures who share similar experiences or emotions can provide a sense of validation and understanding, helping individuals to process and manage their own emotions more effectively. This can be particularly beneficial for those struggling with emotional regulation, as it offers a safe and accessible means of exploring and understanding their emotions. Furthermore, parasocial interaction can provide comfort and distraction during stress or emotional turmoil, helping to alleviate negative emotions and improve overall mental well-being.

Finally, parasocial interaction can contribute to personal growth and self-reflection. By engaging with media figures who challenge our beliefs and perspectives, we are encouraged to examine our own values and assumptions more closely. This process of self-reflection can lead to personal growth and development as we become more aware of our own biases and limitations. Additionally, parasocial interaction can expose us to new ideas and experiences, broadening our horizons and fostering a greater sense of empathy and understanding toward others. In this way, parasocial interaction can contribute to improved mental health outcomes by promoting personal growth and self-awareness.

In conclusion, while parasocial interaction is often viewed with skepticism, it is important to recognize the potential benefits it can offer in terms of mental health. By providing social support, role models, emotional regulation, and opportunities for personal growth, parasocial interaction can improve mental well-being for many individuals. In the next section, we will explore the potential negative consequences of parasocial interaction on mental health and consider strategies for balancing these relationships for optimal mental health outcomes.

The Negative Consequences of Parasocial Interaction on Mental Health

While parasocial interaction can offer several benefits, such as providing a sense of companionship and enhancing self-esteem, it is essential to recognize that it can also have negative consequences on mental health. This section will delve into the potential downsides of parasocial interaction, exploring how excessive engagement with media figures can lead to emotional distress, social isolation, and even addiction-like behaviors.

One of the most significant negative consequences of parasocial interaction is the potential for emotional distress. When individuals become overly attached to media figures, they may experience intense feelings of sadness, disappointment, or even anger when those figures fail to meet their expectations. For example, a fan who has developed a strong parasocial bond with a celebrity may feel devastated if that celebrity becomes involved in a scandal or expresses views that the fan disagrees with. In extreme cases, this emotional distress can lead to symptoms of anxiety and depression, particularly if the individual lacks strong social support networks in real life.

Another potential downside of parasocial interaction is social isolation. When individuals become overly invested in their relationships with media figures, they may neglect their real-life relationships, leading to loneliness and disconnection. This can be particularly problematic for individuals who already struggle with social anxiety or other mental health issues that make it difficult for them to form and maintain healthy relationships. In these cases, parasocial interaction may serve as a maladaptive coping mechanism, allowing individuals to avoid the challenges of real-life social interaction while still experiencing some of the emotional rewards.

In some cases, parasocial interaction can lead to addiction-like behaviors, with individuals increasingly relying on their relationships with media figures to meet their emotional needs. This can manifest in several ways, such as compulsively consuming content related to the media figure, spending excessive amounts of time engaging in online communities dedicated to the figure, or even engaging in risky behaviors to emulate or gain the attention of the figure. These addiction-like behaviors can have serious consequences for an individual's mental health, as they may lead to feelings of guilt, shame, and a sense of losing control over one's life.

Finally, parasocial interaction can lead to unrealistic expectations about real-life relationships. When individuals become accustomed to the idealized, one-sided relationships they form with media figures, they may struggle to navigate the complexities and challenges of real-life relationships. This can result in disappointment, frustration, and even the breakdown of important relationships, as individuals may find it difficult to accept the imperfections and limitations of real-life connections.

In summary, while parasocial interaction can offer some benefits to mental health, it is crucial to recognize the potential negative consequences as well. Emotional distress, social isolation, addiction-like behaviors, and unrealistic expectations about real-life relationships can all result from excessive engagement with media figures. The following section will explore strategies for balancing parasocial interaction to optimize mental health outcomes and mitigate these potential risks.

Balancing Parasocial Interaction for Optimal Mental Health Outcomes

In today's hyper-connected world, parasocial interactions have become integral to our daily lives. While these one-sided relationships can offer numerous benefits, they can also lead to negative consequences if not managed properly. In this section, we will explore the importance of striking a balance between parasocial interaction and real-life relationships to achieve optimal mental health outcomes.

The key to maintaining a healthy balance in parasocial interaction lies in recognizing its potential pitfalls and being mindful of the time and emotional investment dedicated to these relationships.

While it is natural to feel connected to our favorite celebrities or influencers, it is crucial to prioritize our real-life relationships with friends, family, and colleagues. Set boundaries for yourself by allocating specific time slots for engaging in parasocial interactions, such as browsing social media or watching videos. This will help prevent these interactions from encroaching on the time and energy needed to nurture and maintain meaningful connections with the people around you.

Being selective about the content you consume and the platforms you engage with can significantly impact your mental health. Choose to follow individuals who inspire, motivate, and uplift you rather than those who trigger negative emotions or unhealthy comparisons. Additionally, be mindful of the time spent on social media and other platforms, and consider taking periodic breaks to recharge and reconnect with yourself and your real-life relationships.

Developing self-awareness and emotional resilience can help you navigate the complex world of parasocial interactions more effectively. By understanding your emotions and triggers, you can better manage your reactions to the content you consume and the relationships you form. Practice self-reflection and engage in activities that promote emotional well-being, such as journaling, meditation, or therapy.

A strong sense of self is essential for maintaining a healthy balance in parasocial interactions. Focus on building your self-esteem and self-worth by engaging in activities that bring you joy, fulfillment, and a sense of accomplishment. Surround yourself with positive influences and seek opportunities for personal growth and development.

It is essential to be aware of the signs that indicate an unhealthy level of parasocial interaction. These may include excessive time spent on social media, anxiety or depression when not engaging with the object of your parasocial relationship, or neglecting real-life relationships and responsibilities. If you notice these signs, consider seeking professional help or implementing the abovementioned strategies to regain balance in your life.

In conclusion, achieving a healthy balance between parasocial interaction and real-life relationships is crucial for optimal mental health outcomes. By setting boundaries, engaging in mindful consumption of media, cultivating self-awareness and emotional resilience, fostering a healthy sense of self, and recognizing the signs of unhealthy parasocial interaction, individuals can harness the positive aspects of these relationships while mitigating their potential negative consequences. As we continue to navigate the ever-evolving landscape of digital communication, we must remain vigilant and adaptive in our approach to parasocial interaction and mental health.

Strategies for Managing Parasocial Interaction in the Digital Age

In today's fast-paced digital age, parasocial interactions have become increasingly prevalent, thanks to easy access to social media platforms and constant exposure to celebrities and influencers. While these interactions can have both positive and negative effects on mental health, it is crucial to find a balance and manage them effectively. This section will discuss several strategies for managing parasocial interactions in the digital age, ensuring they contribute positively to our mental well-being.

One of the most effective ways to manage parasocial interactions is by setting boundaries and limiting exposure to social media and other platforms where these interactions occur. This can be achieved by allocating specific times of the day for engaging with social media, setting time limits on apps, or even taking periodic breaks from these platforms. By doing so, individuals can prevent excessive engagement in parasocial relationships and maintain a healthy balance between their online and offline lives.

While parasocial interactions can provide a sense of connection and support, it is essential not to neglect real-life relationships. Investing time and energy in nurturing genuine connections with friends, family, and peers can help create a strong support system that contributes to better mental health. Engaging in social activities, joining clubs or organizations, and participating in community events can help foster these connections and reduce the reliance on parasocial relationships for emotional support.

Being mindful of the content we consume and the way we engage with it can significantly impact our mental health. When engaging with celebrities or influencers, one must be aware of the emotions and thoughts that arise during these interactions. By practicing mindfulness, individuals can identify any negative feelings or unhealthy patterns and take steps to address them. This can include unfollowing accounts that trigger negative emotions or seeking professional help if necessary.

In the digital age, it is crucial to develop critical thinking skills to navigate the vast amount of information and content available online. By questioning the authenticity and motives behind the content we consume, we can better understand the impact of parasocial interactions on our mental health. This can also help in distinguishing between genuine connections and those driven by commercial interests or manipulation.

If parasocial interactions begin to negatively impact an individual's mental health or interfere with their daily functioning, it may be necessary to seek professional help. Mental health professionals can provide guidance and support in managing these interactions and addressing any underlying issues that may be contributing to unhealthy engagement in parasocial relationships.

In conclusion, managing parasocial interactions in the digital age is essential for maintaining optimal mental health outcomes. By setting boundaries, cultivating real-life relationships, practicing mindful consumption of media, developing critical thinking skills, and seeking professional help when necessary, individuals can ensure that their engagement in parasocial relationships remains healthy and beneficial. As we continue to navigate the ever-evolving digital landscape, we must remain vigilant and adaptive in our approach to managing parasocial interactions and their impact on our mental well-being.

The Future of Parasocial Interaction and Mental Health

As we have explored throughout this chapter, parasocial interaction is an increasingly relevant and complex phenomenon in today's digital age. With the rapid growth of social media platforms and the constant evolution of technology, it is essential to understand the impact of parasocial interaction on mental health. By examining both the positive and negative consequences, we can better equip ourselves to navigate this ever-changing landscape and promote optimal mental health outcomes.

The future of parasocial interaction is undoubtedly intertwined with the continued advancement of technology and the ways in which we consume media. As virtual reality, augmented reality, and artificial intelligence become more sophisticated and integrated into our daily lives, the lines between parasocial and real-life relationships may become increasingly blurred. This presents both exciting opportunities and potential challenges for mental health professionals, educators, and individuals alike.

In order to harness the positive aspects of parasocial interaction, such as fostering a sense of belonging, providing emotional support, and promoting empathy, it is crucial to develop strategies for managing these relationships in a healthy and balanced manner. This may involve setting boundaries with the amount of time spent engaging in parasocial interactions, cultivating a diverse range of social connections, and being mindful of the potential pitfalls of over-identification with media figures.

Furthermore, mental health professionals and educators must be prepared to address the unique challenges that parasocial interaction can present. This may involve developing targeted interventions and resources to help individuals navigate the complexities of these relationships, as well as promoting media literacy and critical thinking skills to empower individuals to make informed decisions about their engagement with media figures.

Finally, it is important to recognize that the landscape of parasocial interaction will continue to evolve as technology and media consumption habits change. As industry experts, we must remain vigilant and adaptable, continually updating our understanding of this phenomenon and its impact on mental health. By doing so, we can better support individuals in navigating the complexities of parasocial interaction and foster a future where technology and media can be harnessed for the betterment of mental health and well-being.

In conclusion, the future of parasocial interaction and mental health is both exciting and uncertain. However, by understanding the potential benefits and drawbacks of these relationships and by developing strategies to manage them effectively, we can work towards a future where parasocial interaction is a positive force for mental health and well-being. As technology continues to advance and reshape the ways in which we connect with others, it is our responsibility as industry experts to stay informed and adaptable, ensuring that we are prepared to support individuals in navigating this ever-evolving landscape.

Chapter Summary

- Parasocial interaction refers to the one-sided relationships individuals form with media figures, such as celebrities, influencers, or fictional characters, which can have both positive and negative effects on mental health.
- Positive effects of parasocial interaction include providing social support and companionship, serving as role models and inspiration, aiding in emotional regulation and coping, and promoting personal growth and self-reflection.
- Negative consequences of parasocial interaction can include emotional distress, social isolation, addiction-like behaviors, and unrealistic expectations about real-life relationships.
- Balancing parasocial interaction for optimal mental health outcomes involves setting boundaries, engaging in mindful consumption of media, cultivating self-awareness and emotional resilience, fostering a healthy sense of self, and recognizing the signs of unhealthy parasocial interaction.
- Strategies for managing parasocial interaction in the digital age include setting boundaries and limiting exposure, cultivating real-life relationships, practicing mindful media consumption, developing critical thinking skills, and seeking professional help if necessary.

- The future of parasocial interaction is closely tied to the advancement of technology, such as virtual reality, augmented reality, and artificial intelligence, which may blur the lines between parasocial and real-life relationships.
- Mental health professionals and educators must be prepared to address the unique challenges that parasocial interaction can present by developing targeted interventions and resources, and promoting media literacy and critical thinking skills.
- As industry experts, it is crucial to remain vigilant and adaptable, continually updating our understanding of parasocial interaction and its impact on mental health, in order to support individuals in navigating the complexities of these relationships and foster a future where technology and media can be harnessed for the betterment of mental health and well-being.

6

PARASOCIAL INTERACTION IN ONLINE COMMUNITIES AND FANDOMS

An image that represents the emotional bond between a popular Twitch streamer and their dedicated fanbase, showcasing the sense of belonging and friendship within the online community.

In today's digitally connected world, the way we interact with one another has undergone a significant transformation. The rise of social media platforms and online communities has given birth to a new form of social interaction – parasocial interaction. This phenomenon, first coined by Horton and Wohl in 1956, refers to the one-sided relationships that individuals form with media figures, such as celebrities, influencers, and even fictional characters. As we delve into the realm of online communities and fandoms, it becomes crucial to understand the dynamics of parasocial interaction and its implications on our social lives.

Online communities and fandoms are spaces where individuals with shared interests come together to discuss, create, and celebrate their passion for a particular subject, be it a television show, a sports team, or a book series. These virtual spaces provide a sense of belonging and camaraderie, often fostering deep connections between members. However, they also serve as fertile ground for developing parasocial relationships as fans become increasingly invested in the lives and personas of their favorite media figures.

The advent of social media has played a pivotal role in amplifying parasocial interaction. Platforms such as Twitter, Instagram, and YouTube have made it easier than ever for fans to follow, engage with, and feel connected to their favorite celebrities and influencers. These platforms provide a window into the lives of media figures, allowing fans to consume a constant stream of content and updates, further fueling their parasocial attachments.

As we navigate this complex landscape of online communities and fandoms, it is essential to recognize the psychological impact of parasocial interaction on fans and followers. While these relationships can provide a sense of connection and support, they can also lead to unhealthy levels of attachment and even addiction. Understanding the influence of online communities and fandoms on parasocial interaction dynamics is crucial in developing strategies for managing and nurturing healthy parasocial relationships.

In this chapter, we will explore the role of social media in facilitating parasocial relationships, the psychological impact of parasocial interaction on fans and followers, and the influence of online communities and fandoms on parasocial interaction dynamics. We will also consider the future of parasocial interaction in the digital age. By examining these aspects, we aim to provide a comprehensive understanding of parasocial interaction in online communities and fandoms, shedding light on its potential benefits and pitfalls.

The Role of Social Media in Facilitating Parasocial Relationships

In today's digital age, social media has become an integral part of our lives, connecting us with friends, family, and even strangers from around the world. It has also played a significant role in the development and maintenance of parasocial relationships, providing a platform for fans and followers to interact with their favorite celebrities, influencers, and content creators. This section will delve into the ways social media facilitates parasocial relationships and how it has changed the dynamics of these connections.

Before the advent of social media, fans primarily engaged with celebrities through traditional media channels such as television, radio, and print, however, these channels offered limited opportunities for interaction, often leaving fans feeling disconnected from their idols. Social media platforms like Twitter, Instagram, and Facebook have bridged this gap, allowing fans to follow and engage with their favorite personalities in real time. This direct access to celebrities has fostered a sense of intimacy and connection, which is a key component of parasocial relationships.

Social media platforms provide an environment where fans can feel like they are engaging in a reciprocal relationship with their favorite celebrities. This is achieved through features such as likes, comments, and direct messages, which create the illusion of a two-way conversation. Additionally, celebrities often share personal information and behind-the-scenes content, further fostering a sense of intimacy and connection with their fans. This perceived reciprocity and intimacy can strengthen parasocial bonds, making fans feel like they are part of the celebrity's inner circle.

Social media algorithms are crucial in shaping parasocial interactions by curating content based on users' interests and engagement patterns. This can lead to a highly personalized and immersive experience, where fans are constantly exposed to content featuring their favorite celebrities. This constant exposure can intensify parasocial bonds as fans become more invested in the lives of their idols.

As social media has become an essential tool for celebrities to connect with their fans, it has also influenced their behavior and communication styles. Many celebrities have adopted a more authentic and relatable persona on social media, sharing candid moments and personal struggles to create a sense of connection with their fans. This shift in celebrity behavior has further facilitated the development of parasocial relationships, as fans feel they are getting a genuine glimpse into the lives of their favorite personalities.

In conclusion, social media has played a pivotal role in facilitating parasocial relationships by providing a platform for direct interaction between fans and celebrities. The illusion of reciprocity and intimacy, coupled with the impact of algorithms and changes in celebrity behavior, has led to stronger parasocial bonds in the digital age. As social media continues to evolve, it will be interesting to see how these dynamics shift and shape the future of parasocial interaction.

The Psychological Impact of Parasocial Interaction on Fans and Followers

The rise of social media and online communities has led to an increase in parasocial interactions, which are one-sided relationships where an individual invests emotional energy, interest, and time in a media figure, such as a celebrity or influencer, without any reciprocation. These interactions have become an integral part of the fan experience, and understanding their psychological impact is crucial for fans and those they admire. In this section, we will delve into the various ways parasocial interactions can affect fans and followers, both positively and negatively.

One of the most significant psychological impacts of parasocial interactions is the development of emotional attachment and identification with the media figure. Fans often feel a strong connection to their favorite celebrities or influencers, which can lead to a sense of belonging and validation. This emotional attachment can be particularly beneficial for individuals who may feel isolated or marginalized in their offline lives, as it provides a sense of community and support.

Parasocial interactions can also lead to social comparison, where fans compare themselves to the media figures they admire. This can have both positive and negative effects on self-esteem. On the one hand, fans may feel inspired and motivated by the success and achievements of their favorite celebrities, which can boost their self-confidence. On the other hand, constant comparison to seemingly perfect media figures can lead to feelings of inadequacy and low self-esteem, particularly when fans perceive a gap between their own lives and the lives of those they admire.

For many fans, parasocial interactions provide an opportunity for fantasy and escapism. Engaging with the lives of celebrities and influencers can offer a temporary escape from the stresses and challenges of everyday life. While this can be a healthy and enjoyable form of entertainment, it can also become problematic if fans become overly reliant on these parasocial relationships for emotional fulfillment, potentially neglecting their real-life relationships and responsibilities.

In some cases, parasocial interactions can lead to unhealthy levels of obsession and addiction. Fans may become overly invested in the lives of their favorite media figures to the point where their own well-being and relationships suffer. This can manifest in various ways, such as excessive time spent on social media, stalking behaviors, or even harassment of the media figure or other fans. Fans need to recognize the boundaries between healthy admiration and unhealthy obsession, and to seek help if they find themselves struggling with these issues.

Despite the potential negative impacts, parasocial interactions can also catalyze positive change. Fans who feel a strong connection to a media figure may be inspired to adopt their values, beliefs, and behaviors. For example, a fan might be motivated to engage in charitable work, pursue a particular career, or adopt a healthier lifestyle after observing their favorite celebrity's actions. In this way, parasocial interactions can have a significant and lasting impact on fans' lives.

In conclusion, the psychological impact of parasocial interactions on fans and followers is multifaceted and complex. While these relationships can provide a sense of connection, validation, and inspiration, they can also lead to negative outcomes such as low self-esteem, obsession, and addiction. It is crucial for fans to be mindful of the potential risks associated with parasocial interactions and to maintain a healthy balance between their online and offline lives.

The Influence of Online Communities and Fandoms on Parasocial Interaction Dynamics

The advent of the internet and social media platforms has given rise to a new era of communication and connection. Online communities and fandoms have become a significant part of this digital landscape, providing a space for individuals to bond over shared interests, engage in discussions, and form relationships with others who share their passion. In this section, we will explore the ways in which online communities and fandoms influence the dynamics of parasocial interaction, shaping the way fans and followers perceive and interact with their favorite public figures and celebrities.

Online communities, such as forums, fan websites, and social media groups, provide a platform for fans to engage in discussions, share content, and connect with others who share their interests. These communities often serve as a breeding ground for parasocial relationships, as fans are exposed to a constant stream of information and content related to their favorite public figures. This continuous exposure can lead to a deepening sense of familiarity and intimacy with the public figure as fans feel increasingly connected to their lives and experiences.

Moreover, online communities can also foster a sense of collective identity among fans as they bond over their shared admiration for a public figure. This collective identity can further strengthen parasocial relationships, as fans feel a sense of belonging and validation in their connection to the public figure. In turn, this can lead to a heightened emotional investment in the public figure's successes and failures, as fans come to view them as an extension of their own identity.

Fandom culture has long been a driving force behind the formation and maintenance of parasocial relationships. In the digital age, this culture has become even more pervasive, as fans have unprecedented access to their favorite public figures through social media platforms and other online channels. This increased access has led to a shift in the dynamics of parasocial interaction, as fans are now able to engage with public figures in a more direct and personal way.

One notable aspect of fandom culture is the creation and consumption of fan-generated content, such as fan art, fan fiction, and fan videos. This content often serves to further immerse fans in the world of their favorite public figures, deepening their emotional connection and investment in their lives. Additionally, the act of creating and sharing fan-generated content can also serve to reinforce and validate fans' parasocial relationships as they receive feedback and recognition from their fellow fans.

As online communities and fandoms continue to grow and evolve, public figures must also adapt to the changing dynamics of parasocial interaction. Many celebrities and influencers have embraced social media as a way to connect with their fans and foster a sense of intimacy and accessibility. By sharing personal anecdotes, engaging in conversations with fans, and providing behind-the-scenes glimpses into their lives, public figures can strengthen their parasocial relationships and cultivate a loyal and dedicated fanbase.

However, this increased accessibility can also present challenges for public figures, as they must navigate the delicate balance between maintaining their privacy and fostering a sense of connection with their fans. Additionally, public figures must also be mindful of the potential for unhealthy parasocial relationships to develop, as fans may become overly invested in their lives and struggle to separate their online persona from their real-life identity.

In conclusion, online communities and fandoms play a significant role in shaping the dynamics of parasocial interaction in the digital age. By providing a platform for fans to connect with one another and engage with their favorite public figures, these communities can foster a sense of collective identity and emotional investment that strengthens parasocial relationships. As we continue to navigate the ever-evolving digital communication landscape, fans and public figures need to be mindful of the potential pitfalls and challenges associated with parasocial interaction, and strive for healthy and balanced relationships in the online world.

The Future of Parasocial Interaction in the Digital Age

As we have explored throughout this chapter, parasocial interaction has become an increasingly prevalent aspect of our lives in the digital age. Online communities and fandoms have fostered the development of these one-sided relationships, with social media platforms acting as the primary facilitators. The psychological impact of parasocial interaction on fans and followers is complex, and the influence of online communities and fandoms on these dynamics cannot be understated. As we look toward the future, it is essential to consider how parasocial interaction will continue to evolve and what implications this may have for individuals and society as a whole.

One of the most significant factors shaping the future of parasocial interaction is the rapid advancement of technology. Virtual reality, augmented reality, and artificial intelligence are just a few examples of innovations that have the potential to further blur the lines between reality and the digital world. As these technologies become more sophisticated and accessible, it is likely that parasocial relationships will become even more immersive and intense. This raises important questions about the ethical implications of such relationships and the potential for exploitation by creators and influencers.

Another critical aspect to consider is the role of social media platforms in shaping parasocial interaction. As these platforms continue to evolve, they will likely introduce new features and algorithms that further encourage parasocial relationships. For example, the introduction of live streaming and virtual events has allowed fans to feel even closer to their favorite celebrities and influencers. It is crucial for these platforms to balance the desire for increased engagement with the need to protect users' mental health and well-being.

In addition to technological advancements, societal attitudes towards parasocial interaction will also significantly shape its future. As awareness of the phenomenon grows, there may be a shift in how society perceives and values these relationships. This could lead to increased acceptance and understanding of parasocial interaction as a natural part of the human experience or conversely, a backlash against the perceived negative consequences of such relationships.

Finally, it is essential to consider the role of individuals in managing and nurturing healthy parasocial relationships. As we have discussed, there are strategies that fans and followers can employ to ensure that their parasocial interactions remain positive and fulfilling. Encouraging self-awareness, setting boundaries, and engaging in critical thinking are just a few examples of ways in which individuals can take control of their parasocial experiences. It is crucial for fans and followers to recognize the potential risks and benefits of parasocial interaction and to make informed decisions about how they engage with their favorite celebrities and influencers.

In conclusion, the future of parasocial interaction in the digital age is uncertain but undoubtedly fascinating. As technology continues to advance and society grapples with the implications of these one-sided relationships, individuals, creators, and platforms need to work together to ensure that parasocial interaction remains a positive and enriching aspect of our digital lives. By understanding the complexities of parasocial interaction and embracing strategies for healthy engagement, we can navigate the ever-changing landscape of online communities and fandoms with confidence and resilience.

Chapter Summary

- Parasocial interaction refers to one-sided relationships that individuals form with media figures, such as celebrities, influencers, and even fictional characters. Online communities and fandoms provide a space for these relationships to develop.
- Social media platforms play a significant role in facilitating parasocial relationships by providing direct access to celebrities and influencers, creating an illusion of reciprocity and intimacy, and shaping user experiences through algorithms.
- The psychological impact of parasocial interactions on fans and followers can be both positive and negative, including emotional attachment, social comparison, fantasy and escapism, obsession, and addiction, as well as the potential for personal growth and inspiration.
- Online communities and fandoms influence the dynamics of parasocial interaction by fostering a sense of collective identity among fans, promoting fan-generated content, and shaping the behavior of public figures in response to their online presence.
- The rapid advancement of technology, such as virtual reality, augmented reality, and artificial intelligence, has the potential to further blur the lines between reality and the digital world, raising ethical questions about the future of parasocial interaction.
- Social media platforms will continue to evolve, introducing new features and algorithms that further encourage parasocial relationships. Balancing increased engagement with the need to protect users' mental health and well-being is crucial.
- The future of parasocial interaction in the digital age is uncertain but fascinating, with technology, societal

attitudes, and individual responsibility all playing a role in shaping its development. Understanding the complexities of parasocial interaction and embracing strategies for healthy engagement will be essential for navigating the ever-changing landscape of online communities and fandoms.

7

THE DARK SIDE OF PARASOCIAL INTERACTION: OBSESSION AND STALKING

An image depicting a fan's room filled with an overwhelming amount of merchandise, posters, and photos of their favorite celebrity, showcasing the extreme level of obsession in parasocial interaction.

Parasocial interaction refers to the one-sided relationships that individuals form with media figures, such as celebrities, athletes, and fictional characters. These relationships can provide a sense of companionship, emotional support, and even personal identity for those who engage in them. While parasocial interactions can be a harmless and even beneficial aspect of modern media consumption, there is a darker side to this phenomenon that must be acknowledged and addressed.

In this chapter, we will delve into the murky depths of parasocial interaction, exploring the potential dangers that can arise when admiration turns into obsession, and obsession leads to stalking. We will examine the role of social media and online platforms in facilitating these unhealthy behaviors and the psychological impact on both the victims and the perpetrators. Furthermore, we will discuss the legal and ethical implications of parasocial stalking and consider how society can better protect the privacy and mental health of those involved.

As we embark on this exploration, it is essential to recognize that most parasocial interactions are benign and can even be beneficial for individuals seeking connection and solace in an increasingly fragmented world. However, by shining a light on the dark side of parasocial interaction, we can better understand the potential risks and work towards fostering healthier relationships between media consumers and the figures they admire.

In the following sections, we will trace the path from admiration to obsession, highlighting the factors that can contribute to the development of unhealthy parasocial attachments. We will then discuss the role of technology in enabling and exacerbating these behaviors, with a particular focus on social media and online harassment. Next, we will examine the psychological impact of parasocial stalking on both the victims and the perpetrators, and consider the legal and ethical implications of these actions. Finally, we will offer some suggestions for addressing the dark side of parasocial interaction and promoting healthier relationships between media consumers and the figures they admire.

By understanding the potential dangers of parasocial interaction, we can work toward a more responsible and compassionate media landscape where individuals can engage with their favorite figures without crossing the line into obsession and harassment.

The Slippery Slope: From Admiration to Obsession

Parasocial interaction, at its core, is a natural and often harmless part of human behavior. It is the one-sided relationship we form with media figures, such as celebrities, athletes, or fictional characters. However, it can transform into a dark and dangerous obsession when taken to the extreme. This section will explore the slippery slope from admiration to obsession and how seemingly innocent parasocial interactions can morph into something far more sinister.

The journey from admiration to obsession often begins with a deep emotional connection to a media figure. This connection can be fueled by a variety of factors, such as shared interests, perceived similarities, or a sense of relatability. As fans become more invested in the lives of their favorite celebrities, they may begin to feel a sense of ownership or entitlement over them. This can manifest in various ways, such as excessive consumption of content, intrusive thoughts, or even attempts to contact or meet the object of their affection.

One of the key factors that contribute to the development of an unhealthy obsession is the illusion of intimacy. Media figures often share personal details about their lives, making fans feel they have a genuine connection with them. This perceived closeness can lead to an inflated sense of familiarity and a blurring of boundaries between the fan and the celebrity. As a result, fans may begin to believe that they have a special relationship with the media figure, despite the fact that the interaction is entirely one-sided.

The rise of social media has further exacerbated this issue, as it allows fans to have unprecedented access to the lives of their favorite celebrities. Platforms such as Twitter, Instagram, and Facebook provide a constant stream of content and updates, feeding the insatiable appetite of obsessed fans. Furthermore, these platforms offer the illusion of direct communication, as fans can now send messages, leave comments, and even receive "likes" or responses from their favorite media figures. This can reinforce the belief that they have a unique and personal connection, further fueling their obsession.

As the line between admiration and obsession becomes increasingly blurred, fans may begin to exhibit concerning behaviors. These can range from spending excessive amounts of time and money on their favorite celebrity to engaging in cyberstalking or even physical stalking. In some cases, fans may become so consumed by their obsession that it begins to interfere with their daily lives, relationships, and mental health.

It is important to note that not all parasocial interactions will lead to obsession. In fact, most fans are able to maintain a healthy balance between their admiration for a media figure and their own personal lives. However, it is crucial to recognize the potential dangers of parasocial interaction and to be aware of the warning signs of an unhealthy obsession. By understanding the slippery slope from admiration to obsession, we can take steps to foster healthy relationships with media figures and protect ourselves and others from the dark side of parasocial interaction.

Stalking in the Digital Age: Social Media and Online Harassment

In today's hyper-connected world, the internet and social media have become essential tools for communication, self-expression, and staying informed. However, these platforms have also given rise to new forms of stalking and harassment, making it easier for parasocial relationships to escalate into dangerous territory. This section will delve into the various ways that social media and online interactions can contribute to the dark side of parasocial interaction and how these platforms can be both a breeding ground and a weapon for obsession and stalking.

The internet has provided stalkers with unprecedented access to their targets. Social media platforms like Facebook, Instagram, and

Twitter allow users to share personal information, photos, and updates about their lives, often without realizing the potential risks. This wealth of information can be a goldmine for someone with a parasocial obsession, enabling them to monitor their target's every move, gather intimate details about their life, and even manipulate their online environment to create a false sense of closeness.

Online harassment can take many forms, from cyberstalking and doxing (the act of publishing someone's personal information online without their consent) to more subtle tactics like catfishing (creating a fake online persona to deceive someone) and gaslighting (manipulating someone into doubting their own reality). These behaviors can cause significant distress and harm to the target, often leaving them vulnerable, isolated, and powerless.

One of the most insidious aspects of online harassment is its ability to cross the boundary between the virtual and the physical world. For example, a stalker may use information gleaned from social media to locate their target's home or workplace or even to orchestrate "real-life" encounters that can be both terrifying and dangerous. This blurring of lines between the online and offline realms can make it difficult for victims to escape their tormentors and find a safe haven.

The anonymity afforded by the internet can also embolden stalkers, allowing them to hide behind a screen and engage in malicious behavior without fear of being identified or held accountable. This sense of impunity can lead to a vicious cycle of escalation as the stalker becomes increasingly brazen in attempting to control their target.

It is important to recognize that the victims of online harassment and stalking are not the only ones who suffer. The perpetrators may also experience significant psychological distress as their obsession consumes their thoughts and drives them to engage in destructive behavior. In some cases, this may lead to a downward spiral of self-loathing and despair as the stalker becomes trapped in a cycle of compulsion and guilt.

In conclusion, the digital age has brought with it new challenges and dangers when it comes to parasocial interaction. As social media and online platforms continue to evolve, we must remain vigilant in protecting ourselves and others from the dark side of these relationships. By fostering awareness, empathy, and a commitment to ethical online behavior, we can help to create a safer and healthier virtual environment for all.

The Psychological Impact on Victims and Perpetrators

As we delve deeper into the dark side of parasocial interaction, it is crucial to understand its psychological impact on both victims and perpetrators. Obsession and stalking, whether online or offline, can have profound and lasting effects on the mental health of all parties involved.

For victims, the experience of being stalked or harassed can be incredibly distressing. The constant feeling of being watched, monitored, or pursued can lead to perpetual anxiety and fear. This heightened stress can result in various mental health issues, such as depression, panic attacks, and post-traumatic stress disorder (PTSD). Moreover, victims may experience a loss of trust in others and a sense of isolation, as they may feel that no one can truly understand the gravity of their situation.

In addition to the emotional turmoil, victims may also suffer from cognitive and behavioral changes. They may become hyper-vigilant, constantly checking their surroundings and taking extra precautions to ensure their safety. This can lead to sleep disturbances, a decline in work or academic performance, and strained relationships with friends and family. In extreme cases, victims may even feel the need to relocate or change their identity to escape their stalker, further disrupting their lives and sense of stability.

On the other side of the equation, perpetrators of stalking and obsessive behavior also face psychological consequences. While it may be tempting to view stalkers as purely malicious or evil, it is important to recognize that many of these individuals are struggling with their own mental health issues. In some cases, the obsession with a public figure or celebrity may stem from feelings of loneliness, low self-esteem, or a desire for connection and validation.

For some perpetrators, their parasocial relationship with the object of their obsession may provide a temporary escape from their own emotional pain or dissatisfaction with their lives. However, this unhealthy coping mechanism can exacerbate their mental health struggles as they become increasingly disconnected from reality and consumed by their fixation.

Furthermore, stalking or harassing another person can lead to feelings of guilt, shame, and self-loathing for the perpetrator. They may recognize, on some level, that their actions are harmful and inappropriate but feel powerless to stop themselves. This internal conflict can contribute to a cycle of worsening mental health and increasingly destructive behavior.

In conclusion, the psychological impact of obsession and stalking in the context of parasocial interaction is significant and far-reaching, affecting both victims and perpetrators. Society needs to acknowledge and address these mental health consequences in order to better support those who have been affected and to prevent future instances of harmful parasocial behavior. By fostering open conversations and promoting healthy relationships, we can work towards a more empathetic and understanding world.

Legal and Ethical Implications: Protecting Privacy and Mental Health

As we delve deeper into the dark side of parasocial interaction, it becomes crucial to address the legal and ethical implications surrounding obsession and stalking. In this section, we will explore the importance of protecting privacy and mental health for both the victims and the perpetrators, as well as the challenges faced by lawmakers and society in dealing with these issues.

One of the primary legal concerns in cases of parasocial obsession and stalking is the right to privacy. Celebrities, influencers, and public figures often have to strike a delicate balance between maintaining their public image and protecting their personal lives from unwanted intrusion. While fans may argue that they have a right to know about the lives of their idols, it is essential to recognize that everyone, regardless of their public status, is entitled to a certain level of privacy.

Laws surrounding privacy vary from country to country, but most jurisdictions have some form of legal protection against stalking and harassment. However, the enforcement of these laws can be challenging, particularly in the digital age, where perpetrators can easily hide behind anonymity and cross borders with a few clicks.

The mental health of both victims and perpetrators is another critical aspect to consider when discussing the legal and ethical implications of parasocial obsession and stalking. Victims often experience severe emotional distress, anxiety, and even post-traumatic stress disorder (PTSD) as a result of being stalked or harassed. In some cases, the psychological impact can be so severe that it affects their ability to work, maintain relationships, and lead a normal life.

On the other hand, perpetrators may also struggle with mental health issues, such as personality disorders, attachment issues, or delusional beliefs, which can contribute to their obsessive behavior. While it is crucial to hold perpetrators accountable for their actions, it is also essential to recognize the need for mental health support and intervention to prevent further harm to themselves and others.

Social media platforms and online communities play a significant role in facilitating parasocial interaction, and they also bear some responsibility for addressing the dark side of these relationships. Platforms like Facebook, Instagram, and Twitter have implemented policies and reporting tools to combat harassment and stalking. However, there is still much work to be done in terms of improving these systems and ensuring that they are effective in protecting users' privacy and mental health.

Moreover, online communities, such as fan forums and discussion groups, should also strive to create a safe and supportive environment for their members. This includes establishing clear guidelines for acceptable behavior, promoting healthy boundaries, and providing resources for those struggling with obsession or stalking issues.

In conclusion, addressing the legal and ethical implications of parasocial obsession and stalking requires a collective effort from various stakeholders, including lawmakers, social media platforms, online communities, mental health professionals, and the public at large. By working together, we can foster a greater understanding of the complexities surrounding parasocial interaction, protect the privacy and mental health of those involved, and ultimately promote healthier relationships between fans and public figures.

Addressing the Dark Side and Fostering Healthy Relationships

As we have explored throughout this chapter, parasocial interactions can take a dark turn, leading to obsession and stalking. While these behaviors are not the norm, they are important to address and understand in order to foster healthier relationships with public figures and celebrities. In this conclusion, we will summarize the key points discussed in this chapter and provide recommendations for addressing the dark side of parasocial interactions and promoting healthier connections.

Firstly, it is crucial to recognize the slippery slope from admiration to obsession. While it is natural to feel a connection with public figures, it is important to maintain a sense of perspective and boundaries. This can be achieved by reminding oneself of the one-sided nature of parasocial relationships and the importance of cultivating real-life connections with friends and family.

Secondly, the digital age has made stalking and online harassment easier than ever before. Social media platforms should take responsibility for creating a safe environment for both public figures and users. This can be done by implementing stricter policies against harassment, providing resources for users to report and block abusive behavior, and promoting digital literacy and online safety education.

Thirdly, understanding the psychological impact on both victims and perpetrators is essential in addressing the dark side of parasocial interactions. Providing support and resources for victims of stalking and harassment is crucial, as well as offering intervention and rehabilitation programs for perpetrators. Mental health professionals can play a vital role in helping individuals navigate the complexities of parasocial relationships and their potential negative consequences.

Fourthly, legal and ethical implications must be considered in order to protect privacy and mental health. This includes updating and enforcing laws that address stalking and harassment, both online and offline. Additionally, public figures and celebrities should be encouraged to speak out about their experiences and advocate for stronger protection and support systems.

Lastly, fostering healthy relationships with public figures and celebrities involves promoting a culture of empathy, respect, and understanding. This can be achieved through education and awareness campaigns that highlight the importance of maintaining boundaries and respecting the privacy of others. Encouraging open conversations about the potential pitfalls of parasocial interactions can help individuals develop a more balanced and realistic perspective on their relationships with public figures.

In conclusion, addressing the dark side of parasocial interactions is a multifaceted challenge that requires the cooperation of individuals, mental health professionals, social media platforms, and legal systems. By acknowledging the potential dangers and promoting healthy relationships, we can work towards a society where parasocial interactions are a source of inspiration and connection rather than obsession and harm.

Chapter Summary

- Parasocial interactions refer to one-sided relationships formed with media figures, which can provide companionship and emotional support. However, there is a dark side to this phenomenon, including obsession and stalking.
- The slippery slope from admiration to obsession often begins with a deep emotional connection to a media figure, which can be fueled by shared interests, perceived

similarities, or a sense of relatability. The illusion of intimacy can lead to an inflated sense of familiarity and a blurring of boundaries.

- Social media and online platforms have exacerbated the issue of parasocial obsession and stalking, providing unprecedented access to the lives of public figures and enabling intrusive behavior.
- The psychological impact of parasocial stalking is significant for both victims and perpetrators. Victims may experience anxiety, depression, and PTSD, while perpetrators may struggle with loneliness, low self-esteem, and a desire for connection.
- Legal and ethical implications surrounding parasocial obsession and stalking include balancing privacy rights and public interest, addressing the mental health of both victims and perpetrators, and the role of social media platforms and online communities in facilitating these behaviors.
- Protecting privacy and mental health is crucial in addressing the dark side of parasocial interactions. This involves updating and enforcing laws against stalking and harassment, as well as providing support and resources for victims and intervention programs for perpetrators.
- Fostering healthy relationships with public figures involves promoting a culture of empathy, respect, and understanding. Education and awareness campaigns can help individuals maintain boundaries and respect the privacy of others.
- Addressing the dark side of parasocial interactions requires the cooperation of individuals, mental health professionals, social media platforms, and legal systems. By acknowledging the potential dangers and promoting healthy relationships, society can work towards a more responsible and compassionate media landscape.

8

NAVIGATING PARASOCIAL RELATIONSHIPS: BOUNDARIES AND ETHICS

An image that represents the emotional bond between a popular Twitch streamer and their dedicated fanbase, showcasing the sense of belonging and friendship within the online community.

In today's digital age, the way we connect with others has evolved dramatically. The rise of social media platforms and the internet has given birth to a new form of interaction – parasocial relationships. These one-sided relationships have become increasingly prevalent in our society, where an individual feels a strong emotional connection to a public figure or influencer without direct interaction. As we continue to navigate this ever-changing landscape, it is crucial to understand the importance of parasocial relationships and their impact on individuals and society.

Parasocial relationships have existed for decades, dating back to the early days of television and radio. However, the advent of social media has amplified their presence and influence in our lives. With just a few clicks, we can access the lives of our favorite celebrities, influencers, and content creators, fostering a sense of intimacy and connection that may not have been possible in the past. This accessibility has led to a blurring of lines between public and private lives, making it essential for us to examine the nature of these relationships and their potential consequences.

Understanding parasocial relationships is vital for several reasons. First, it allows us to recognize the emotional investment that individuals may have in these connections. For some, parasocial relationships can provide a sense of comfort, belonging, and even inspiration. However, for others, these relationships may lead to unhealthy attachment, obsession, or even stalking behaviors. By examining the dynamics of parasocial relationships, we can better identify the warning signs of unhealthy connections and provide support and resources for those who may be struggling.

Second, understanding parasocial relationships is crucial for content creators and influencers themselves. As public figures, they have a responsibility to be aware of the impact their actions may have on their audience. By acknowledging the power dynamics at play, they can make informed decisions about how to engage with their followers and create content that fosters positive connections while minimizing potential harm.

Finally, social media platforms play a significant role in shaping parasocial relationships. By understanding the mechanics of these connections, platforms can develop features and policies that promote responsible engagement and discourage negative behaviors. This can lead to a safer and more enjoyable online environment for all users.

In this chapter, we will delve deeper into the world of parasocial relationships, exploring the importance of establishing healthy boundaries, ethical considerations for content creators and influencers, and the role of social media platforms in fostering responsible parasocial interactions. Additionally, we will discuss strategies for managing and mitigating negative parasocial behaviors and consider the future of parasocial relationships and their impact on society.

Establishing Healthy Boundaries in Parasocial Interactions

As we delve into the world of parasocial interactions, it is crucial to understand the importance of establishing healthy boundaries between content creators, influencers, and their audiences. These boundaries protect both parties' mental and emotional well-being and ensure that the relationship remains positive and mutually beneficial. This section will explore various strategies for setting and maintaining appropriate boundaries in parasocial interactions.

The first step in establishing healthy boundaries is acknowledging that parasocial relationships, by their very nature, are limited in scope and depth. Unlike traditional friendships or romantic relationships, parasocial interactions are primarily one-sided, with the audience investing emotional energy into the relationship without receiving direct reciprocation from the content creator or influencer. Both parties need to recognize these limitations and adjust their expectations accordingly.

Content creators and influencers must be proactive in setting personal boundaries to protect their privacy and well-being. This may include:

- Limiting the amount of personal information shared with the audience, such as details about family, relationships, or home life.
- Establishing clear guidelines for communication, such as designated times for responding to messages or comments and specifying which topics are off-limits.
- Taking regular breaks from social media and online interactions to recharge and maintain a healthy work-life balance.

While content creators and influencers play a significant role in setting boundaries, it is also essential for audiences to engage in healthy and respectful ways. This include:

- Respecting the privacy and personal boundaries of content creators and influencers, refraining from intrusive or inappropriate questions, and avoiding overstepping into their personal lives.
- Recognizing that content creators and influencers are not obligated to respond to every message or comment and respecting their need for personal time and space.
- Engaging in positive and constructive discussions, avoiding toxic or harmful behaviors, and promoting a supportive online community.

Social media platforms and online communities should also take responsibility for fostering healthy parasocial interactions in the following ways:

- Implementing clear guidelines and policies to protect the privacy and well-being of content creators, influencers, and their audiences.
- Providing tools and resources for users to manage their online interactions, such as privacy settings, content filters, and options for reporting inappropriate behavior.

- Encouraging a culture of respect and empathy within online communities, promoting positive engagement, and discouraging harmful or toxic behaviors.

By understanding the nature of parasocial relationships and taking proactive steps to establish and maintain healthy boundaries, content creators, influencers, and their audiences can enjoy the benefits of these unique connections while minimizing potential risks and negative consequences.

Ethical Considerations for Content Creators and Influencers

As parasocial relationships continue to thrive in the digital age, content creators and influencers must be aware of the ethical considerations that come with their unique position in the lives of their followers. These individuals often hold significant power over their audience, and with that power comes a responsibility to act ethically and conscientiously. In this section, we will discuss several key ethical considerations that content creators and influencers should keep in mind as they navigate the complex world of parasocial relationships.

One of the most important ethical considerations for content creators and influencers is the need for transparency and authenticity in their interactions with their audience. This means being open and honest about sponsored content, partnerships, and any other potential conflicts of interest that may arise. By being transparent, content creators can maintain the trust of their audience and ensure that their parasocial relationships are built on a foundation of honesty and integrity.

Another crucial ethical consideration for content creators and influencers is respecting the privacy and personal boundaries of their audience. This includes being mindful of the information they share about their followers, as well as being sensitive to the potential impact of their content on the mental and emotional well-being of their audience. Content creators should strive to create a safe and inclusive space for their followers to engage in meaningful parasocial interactions without fear of exploitation or harm.

As content creators and influencers hold significant power over their audience, it is essential that they avoid using this power to manipulate or exploit their followers. This means being mindful of the potential impact of their content and actions on their audience, and ensuring that they do not take advantage of their followers' trust or vulnerability. Content creators should be especially cautious when promoting products or services, as their influence can significantly impact the purchasing decisions of their audience.

Content creators and influencers have a unique opportunity to shape the behavior and attitudes of their followers. As such, they should strive to promote positive and responsible behavior in their content and interactions with their audience. This includes being mindful of the messages they send through their content and actively encouraging their followers to engage in healthy and responsible parasocial interactions.

Finally, content creators and influencers should recognize the importance of promoting critical thinking and media literacy among their audience. By encouraging their followers to question and analyze their content, content creators can help foster a more informed and discerning audience, better equipped to navigate the complex world of parasocial relationships.

In conclusion, content creators and influencers play a significant role in shaping the parasocial relationships of their audience. By being mindful of the ethical considerations discussed in this section, they can help to ensure that these relationships are built on a foundation of trust, respect, and mutual understanding, ultimately contributing to a healthier and more responsible digital landscape.

The Role of Social Media Platforms in Fostering Responsible Parasocial Relationships

In today's digital age, social media platforms have become the primary medium for parasocial interactions, allowing fans to feel a sense of connection with their favorite content creators and influencers. As these platforms continue to grow and evolve, it is crucial to examine their role in fostering responsible parasocial relationships and ensuring that both creators and their audiences engage in healthy, ethical interactions.

Social media algorithms play a significant role in shaping the parasocial landscape. By prioritizing content that generates high levels of engagement, these algorithms can inadvertently encourage creators to cultivate parasocial relationships with their audience. While this can lead to increased visibility and success for creators, it also raises concerns about the potential for manipulation and exploitation of fans' emotions.

To foster responsible parasocial relationships, social media platforms must be transparent about their algorithms and work to minimize the potential for harmful consequences. This could involve adjusting algorithms to prioritize content that promotes healthy interactions and discourages manipulative tactics, as well as providing users with tools to better understand and control their own engagement with content.

One of the most effective ways to promote responsible parasocial relationships is by fostering digital literacy and media education among users. Social media platforms can play a key role in this process by providing resources and tools that help users understand the nature of parasocial interactions and their potential risks.

This could include educational content about the psychological aspects of parasocial relationships, tips for setting healthy boundaries, and guidelines for ethical behavior in online interactions. Social media platforms can help create a more informed and responsible online community by empowering users with knowledge and resources.

Social media platforms must also take responsibility for the content shared on their platforms by implementing clear policies and guidelines for creators. These guidelines should address issues such as harassment, manipulation, and the sharing of misleading or harmful information.

By holding creators accountable for their actions and promoting ethical behavior, social media platforms can help to create a safer and more responsible environment for parasocial interactions. This may also involve providing support and resources for creators to better understand and navigate the complexities of parasocial relationships, ensuring that they are equipped to engage with their audience in a healthy and ethical manner.

Lastly, social media platforms can foster responsible parasocial relationships by providing users with tools to manage their online interactions. This could include features that allow users to set boundaries on their engagement with specific creators, limit the amount of time spent on the platform, or filter content based on their personal preferences and values.

By giving users more control over their online experience, social media platforms can empower them to make informed decisions about their parasocial relationships and minimize the potential for negative consequences.

In conclusion, social media platforms have a significant role to play in fostering responsible parasocial relationships. By addressing the power of algorithms, promoting digital literacy, implementing clear policies for creators, and providing tools for users to manage their online interactions, these platforms can help to create a healthier and more ethical environment for parasocial interactions. As we continue to navigate the ever-evolving digital landscape, we must prioritize the well-being of both creators and their audiences, ensuring that parasocial relationships remain a positive and enriching aspect of our online experiences.

Strategies for Managing and Mitigating Negative Parasocial Behaviors

As we delve deeper into the world of parasocial relationships, it becomes increasingly important to address the potential negative aspects that may arise from these interactions. While parasocial relationships can foster a sense of connection and community, they can also lead to unhealthy behaviors and unrealistic expectations. In this section, we will explore various strategies for managing and mitigating negative parasocial behaviors, both for individuals engaging in these relationships and for content creators and influencers.

- **Self-awareness and reflection:** The first step in managing negative parasocial behaviors is to recognize them. Individuals should take the time to reflect on their relationships with public figures and assess whether their level of emotional investment is healthy and sustainable. It is essential to maintain a balance between one's personal life and the time and energy dedicated to following and engaging with public figures.
- **Establishing boundaries:** As discussed in the previous section, setting healthy boundaries is crucial in maintaining a positive parasocial relationship. This includes limiting the time spent consuming content, engaging in online discussions, or fantasizing about the public figure. It is also important to respect the privacy of the public figure and refrain from intrusive or obsessive behaviors.
- **Diversifying interests and social connections:** Relying solely on parasocial relationships for emotional support and social interaction can be detrimental to one's mental health. It is essential to cultivate diverse interests and maintain connections with friends, family, and peers in the real world. This can help prevent over-reliance on parasocial relationships and reduce the risk of negative behaviors.
- **Seeking professional help:** If an individual struggles to manage their parasocial behaviors or experiences significant distress as a result, it may be beneficial to seek professional help. Mental health professionals can provide guidance and support in addressing unhealthy patterns and developing coping strategies.

For content creators and influencers, there are several strategies to consider in mitigating negative parasocial behaviors among their audience:

- **Promoting healthy engagement:** Content creators can set the tone for their audience by encouraging respectful and responsible interactions. This may include addressing the issue of parasocial relationships directly, reminding followers of the importance of boundaries, and discouraging obsessive or harmful behaviors.
- **Modeling self-care and balance:** Public figures can help normalize the importance of self-care and maintaining a balanced life by sharing their own experiences and strategies. This can serve as a reminder to their audience that it is essential to prioritize their own well-being and personal relationships.
- **Collaborating with mental health professionals:** Content creators can partner with mental health professionals to provide resources and information on managing parasocial relationships. This can help raise awareness of the potential risks and encourage responsible engagement among their audience.
- **Monitoring and addressing negative behaviors:** Content creators should be vigilant in monitoring their online communities for signs of unhealthy parasocial behaviors. This may involve moderating comments, addressing concerning behaviors directly, or even seeking professional guidance on how to handle specific situations.

In conclusion, managing and mitigating negative parasocial behaviors is a shared responsibility between individuals engaging in these relationships and the public figures they admire. By promoting self-awareness, establishing boundaries, and fostering healthy engagement, we can work towards a future where parasocial relationships serve as a positive and enriching aspect of our increasingly interconnected world.

The Future of Parasocial Relationships and Their Impact on Society

As we reach the end of our exploration into the complex world of parasocial interactions, it is essential to consider the future of these relationships and their potential impact on society. With the rapid growth of technology and the increasing influence of social media, it is clear that parasocial relationships will continue to evolve and shape the way we interact with one another.

One of the most significant aspects of this evolution is the increasing accessibility of content creators and influencers. As the barriers between celebrities and their audiences continue to blur, both parties must maintain a healthy understanding of the boundaries that should exist in these relationships. By fostering a culture of respect and responsibility, we can ensure that parasocial interactions remain a positive force in our lives.

Moreover, as the line between reality and virtual reality becomes increasingly blurred, it is essential to consider the ethical implications of these relationships. Content creators and influencers must be aware of the power they hold over their audiences and use it responsibly. This includes being transparent about sponsored content, avoiding the promotion of harmful behaviors, and being mindful of the impact their actions may have on their followers.

Social media platforms also have a role to play in fostering responsible parasocial relationships. By implementing policies and features that promote transparency and discourage harmful behaviors, these platforms can help create a safer environment for both creators and their audiences. This may include features such as clear labeling of sponsored content, tools to report and address harassment, and educational resources for users to better understand the nature of parasocial interactions.

As we move forward, it is crucial for individuals to develop strategies for managing and mitigating negative parasocial behaviors. This may involve setting personal boundaries, seeking support from friends and family, or engaging in self-reflection to better understand one's own motivations and expectations in these relationships. By taking a proactive approach, we can minimize the potential harm that may arise from unhealthy parasocial interactions.

In conclusion, the future of parasocial relationships is both exciting and uncertain. As technology continues to advance and our world becomes increasingly interconnected, society needs to adapt and grow alongside these changes. By fostering a culture of responsibility, ethical behavior, and self-awareness, we can ensure that parasocial relationships remain a positive and enriching aspect of our lives. Ultimately, the key to navigating this new frontier lies in our ability to strike a balance between the benefits of these relationships and the potential risks they may pose.

Chapter Summary

- Parasocial relationships, or one-sided emotional connections with public figures, have become increasingly prevalent in today's digital age, making it crucial to understand their impact on individuals and society.
- Establishing healthy boundaries in parasocial interactions is essential for protecting the mental and emotional well-being of both content creators and their audiences.
- Content creators and influencers have a responsibility to act ethically and conscientiously, including being transparent, respecting privacy, avoiding manipulation, promoting positive behavior, and encouraging critical thinking.
- Social media platforms play a significant role in shaping parasocial relationships and should prioritize responsible engagement, promote digital literacy, implement clear policies for creators, and provide tools for users to manage their online interactions.
- Recognizing the limits of parasocial relationships and adjusting expectations accordingly is crucial for maintaining a healthy emotional investment in these connections.

- Strategies for managing and mitigating negative parasocial behaviors include self-awareness, establishing boundaries, diversifying interests and social connections, and seeking professional help if necessary.
- Content creators can help mitigate negative parasocial behaviors among their audience by promoting healthy engagement, modeling self-care, collaborating with mental health professionals, and monitoring and addressing concerning behaviors.
- The future of parasocial relationships depends on fostering a culture of responsibility, ethical behavior, and self-awareness, ensuring that these connections remain a positive and enriching aspect of our increasingly interconnected world.

9

THE FUTURE OF PARASOCIAL INTERACTION: VIRTUAL REALITY AND ARTIFICIAL INTELLIGENCE

An image depicting a person engaging in a deep conversation with their favorite virtual influencer, showcasing the blurred lines between reality and the digital world in parasocial interactions.

As we stand on the cusp of a new era in human communication and interaction, it is essential to explore the ever-evolving landscape of parasocial relationships. These one-sided connections, where an individual feels a bond with a media figure or character, have long been a staple of our media consumption. From the early days of radio and television to the rise of social media influencers, parasocial interactions have continued to grow in complexity and significance. Now, with the advent of virtual reality (VR) and artificial intelligence (AI), we are poised to witness a paradigm shift in the way we form and maintain these relationships.

This chapter will delve into the exciting and uncharted territory of parasocial interaction in the digital age. We will examine the potential of VR and AI to create immersive experiences and authentic virtual personalities, fostering deeper and more meaningful connections between individuals and their virtual counterparts. Furthermore, we will discuss the ethical considerations that arise as we balance the benefits of technological advancements with the need to preserve the genuine human connection. Finally, we will explore the potential impacts of these innovations on the media and entertainment industries, as well as society as a whole.

As we embark on this journey, it is crucial to approach these emerging technologies with a sense of curiosity and open-mindedness. The future of parasocial interaction is brimming with possibilities, and it is up to us to navigate these uncharted waters with wisdom and foresight. By understanding the potential of VR and AI to transform our parasocial experiences, we can harness their power to create a more connected, empathetic, and engaging world.

Virtual Reality: Immersive Experiences and Enhanced Parasocial Relationships

The advent of virtual reality (VR) technology has opened up a new frontier in the realm of parasocial interaction. By providing users with immersive experiences that blur the lines between the real and the virtual, VR has the potential to significantly enhance the depth and intensity of parasocial relationships. In this section, we will explore the ways in which virtual reality is revolutionizing the way we engage with media figures and fictional characters, as well as the implications of these developments for the future of parasocial interaction.

One of the most striking features of virtual reality is its ability to create a sense of presence or the feeling of being physically located within a virtual environment. This heightened sense of immersion allows users to experience a more direct and intimate connection with the virtual world and its inhabitants. In the context of parasocial interaction, this means that users can now interact with their favorite media figures and characters in a way that feels more real and immediate than ever before.

For instance, imagine watching a concert through a VR headset, where you can see and hear the performance and feel like you are standing in the front row, surrounded by other fans. This level of immersion can lead to a stronger emotional bond between the user and the performer, as the virtual experience closely mimics the sensations of attending a live event. Similarly, VR gaming allows players to step into the shoes of their favorite characters, experiencing their adventures firsthand and forging a deeper connection with the virtual world.

Another aspect of virtual reality that has significant implications for parasocial interaction is the potential for interactivity. Many VR experiences are designed to be interactive, allowing users to engage with the virtual environment and its inhabitants more actively and dynamically. This interactivity can lead to a greater sense of agency and control, as users are no longer passive observers but active participants in the unfolding narrative.

For example, in a VR-based talk show, users might be able to ask questions directly to the host or guests, or even participate in the conversation themselves. This level of engagement can foster a stronger sense of connection and identification with the media figures involved, as users feel that they are part of the conversation and that their input is valued.

However, it is important to note that the increased immersion and interactivity offered by virtual reality can also lead to potential pitfalls in the realm of parasocial interaction. As users become more deeply engaged with virtual personalities, there is a risk that the boundaries between the real and the virtual may become blurred, leading to unhealthy levels of attachment or even addiction. Furthermore, the ethical implications of creating increasingly realistic and responsive virtual beings must be carefully considered as we navigate the uncharted waters of parasocial interaction in the digital age.

In conclusion, virtual reality has the potential to significantly enhance the depth and intensity of parasocial relationships, offering users immersive and interactive experiences that were previously unimaginable. However, as we embrace this new era of parasocial interaction, we must remain mindful of the potential risks and ethical considerations that accompany these technological advancements. By striking a balance between innovation and responsibility, we can ensure that virtual reality is a powerful tool for fostering meaningful connections and enriching our understanding of the human experience.

Artificial Intelligence: Crafting Authentic and Responsive Virtual Personalities

As we delve deeper into parasocial interaction, it is crucial to explore the role of artificial intelligence (AI) in crafting authentic and responsive virtual personalities. AI has the potential to revolutionize the way we interact with our favorite characters and celebrities, creating a more immersive and personalized experience for audiences.

The development of AI technology has made significant strides in recent years, enabling the creation of virtual personalities that can engage with users in a more natural and human-like manner. These AI-driven characters can analyze and respond to user inputs, adapting their behavior and responses to create a more dynamic and interactive experience. This level of responsiveness allows for a deeper level of parasocial interaction, as users can develop a stronger emotional connection with virtual personalities that seem to understand and respond to their needs and desires.

One of the key aspects of crafting authentic virtual personalities is the ability to generate realistic and engaging dialogues. Natural language processing (NLP) and machine learning algorithms enable AI-driven characters to understand and respond to user inputs in a more human-like manner. This not only enhances the user's experience but also contributes to the development of a more genuine parasocial relationship.

Another important factor in creating authentic virtual personalities is the development of emotional intelligence in AI systems. By incorporating emotional recognition and response capabilities, AI-driven characters can better understand and react to the user's emotional state, providing a more empathetic and supportive interaction. This can lead to a stronger emotional bond between the user and the virtual personality, further blurring the line between reality and the virtual world.

As AI technology continues to advance, we can expect to see even more sophisticated virtual personalities that can engage with users on a deeper level. These AI-driven characters may be able to learn and adapt to individual users, tailoring their interactions to create a truly personalized experience. This level of customization could lead to the development of virtual companions, mentors, or even therapists, providing users with unique and valuable resource for personal growth and emotional support.

However, as we embrace the potential of AI-driven virtual personalities, it is essential to consider the ethical implications of these advancements. While the prospect of more authentic and responsive virtual personalities is undoubtedly exciting, we must also be mindful of the potential consequences of relying too heavily on AI-driven interactions. As we continue exploring the future of parasocial interaction, it is crucial to balance the benefits of technological advancements and the importance of maintaining genuine human connections.

In conclusion, artificial intelligence has the potential to revolutionize the way we experience parasocial interaction, creating more authentic and responsive virtual personalities that can engage with users on a deeper level. As we navigate the uncharted waters of parasocial interaction in the digital age, it is essential to consider both the exciting possibilities and the ethical challenges that come with these advancements. By doing so, we can ensure that we harness the power of AI to enhance our parasocial experiences while maintaining a healthy balance between the virtual and the real world.

Ethical Considerations: Balancing Technological Advancements and Human Connection

As we venture into the uncharted waters of parasocial interaction in the digital age, it is crucial to address the ethical considerations that arise with the integration of virtual reality (VR) and artificial intelligence (AI) into our daily lives. The rapid advancements in these technologies have the potential to significantly impact the way we form and maintain relationships, both with virtual entities and with other human beings. In this section, we will explore the ethical challenges that must be addressed in order to strike a balance between embracing technological advancements and preserving genuine human connection.

One of the primary ethical concerns surrounding parasocial interaction in the digital age is the authenticity of virtual relationships. As AI-powered virtual entities become increasingly sophisticated and responsive, it becomes more difficult to distinguish between genuine human interaction and interactions with virtual beings. This raises questions about the value and authenticity of these relationships and the potential for deception and manipulation.

To address this concern, it is essential to establish clear guidelines and ethical standards for developing and using AI-driven virtual entities. This may include transparency about the nature of these virtual beings, as well as limitations on their ability to mimic human emotions and behaviors. By doing so, we can ensure that users are able to make informed decisions about the relationships they choose to engage in and maintain a clear distinction between virtual and real-world connections.

Another ethical consideration in parasocial interaction is the issue of privacy and data security. As users engage with virtual entities through VR and AI platforms, they may inadvertently share sensitive personal information or expose themselves to potential data breaches. This raises concerns about the potential misuse of personal data and the potential for virtual entities to be used as tools for surveillance or manipulation.

To mitigate these risks, it is crucial to establish robust data protection measures and privacy policies for VR and AI platforms. This may include end-to-end encryption, strict access controls, and transparent data collection practices. Additionally, users should be educated about the potential risks associated with sharing personal information in virtual environments and encouraged to take proactive steps to protect their privacy.

As virtual reality and artificial intelligence continue to advance, there is a risk that individuals may become increasingly reliant on virtual relationships at the expense of real-world connections. This could potentially lead to social isolation and a decline in the quality of human relationships as individuals prioritize virtual interactions over genuine human connection.

To address this concern, it is important to promote a balanced approach to the use of VR and AI technologies. This may involve encouraging users to engage in a mix of virtual and real-world social activities and fostering a culture of digital well-being that emphasizes the importance of maintaining strong human connections. By doing so, we can ensure that the future of parasocial interaction is one that enhances, rather than detracts from, our ability to connect with one another on a deeper level.

In conclusion, the ethical considerations surrounding the future of parasocial interaction in the digital age are complex and multifaceted. By addressing these concerns head-on and establishing clear guidelines and ethical standards for developing and using VR and AI technologies, we can strike a balance between embracing the potential of these advancements and preserving the genuine human connection that lies at the heart of our social experience.

Potential Impacts: How VR and AI Will Shape the Future of Media and Entertainment

As we venture further into the digital age, the integration of virtual reality (VR) and artificial intelligence (AI) in media and entertainment is inevitable. These technologies have the potential to revolutionize the way we consume content, interact with virtual characters, and form parasocial relationships. In this section, we will explore the potential impacts of VR and AI on the future of media and entertainment, and how these advancements will redefine parasocial interaction.

One of the most significant impacts of VR and AI on media and entertainment is the ability to create personalized and interactive content. With AI algorithms, content creators can analyze user preferences, behaviors, and emotions to tailor experiences that cater to individual tastes. This personalization can lead to stronger parasocial relationships, as audiences feel a deeper connection to the virtual characters and stories that resonate with their interests and values.

Moreover, VR technology allows users to immerse themselves in interactive environments where they can actively participate in the narrative and influence its outcome. This level of interactivity can foster a sense of agency and emotional investment in the virtual world, further strengthening parasocial bonds.

As AI continues to advance, we can expect to see a surge in the popularity of virtual influencers and celebrities. These digital personas, powered by AI, can engage with their audience in real-time, respond to their needs and adapt their behavior accordingly. This responsiveness can lead to more authentic and dynamic parasocial relationships as virtual characters become increasingly indistinguishable from their human counterparts.

Furthermore, virtual celebrities can exist indefinitely, allowing for long-lasting parasocial relationships that transcend the limitations of human lifespan and availability. This permanence can create a sense of stability and familiarity for audiences, who can continue to engage with their favorite virtual personalities over time.

The integration of VR and AI in media and entertainment also raises several ethical concerns. As virtual characters become more realistic and responsive, there is a risk of users becoming overly attached to these digital personas, potentially leading to social isolation and a preference for virtual relationships over real-life connections.

Moreover, the use of AI algorithms to create personalized content can result in echo chambers, where users are only exposed to information and perspectives that align with their existing beliefs. This can limit critical thinking and perpetuate biases, ultimately hindering personal growth and societal progress.

As a result, there is a pressing need for regulation and ethical guidelines to ensure that the use of VR and AI in media and entertainment promotes healthy parasocial relationships and fosters a diverse and inclusive virtual landscape.

In conclusion, the future of parasocial interaction in the digital age is both exciting and uncertain. The integration of VR and AI in media and entertainment has the potential to enhance parasocial relationships, create more engaging and personalized content, and give rise to a new generation of virtual influencers and celebrities. However, these advancements also come with ethical challenges that must be addressed to ensure a balanced and responsible approach to technology. As we navigate the uncharted waters of parasocial interaction in the digital age, it is crucial to remain vigilant and adaptive, embracing the potential of VR and AI while safeguarding the essence of human connection.

Navigating the Uncharted Waters of Parasocial Interaction in the Digital Age

As we stand on the precipice of a new era in parasocial interaction, it is crucial to recognize the potential opportunities and challenges that lie ahead. The integration of virtual reality and artificial intelligence into our media landscape has the power to revolutionize the way we form connections with virtual personalities, blurring the lines between reality and fiction like never before. However, with great power comes great responsibility, and it is up to us as a society to navigate these uncharted waters with care and consideration.

The advent of virtual reality has opened the door to immersive experiences that allow us to engage with virtual personalities on a deeper level. By stepping into their worlds and interacting with them in a more tangible way, we can form stronger parasocial bonds that transcend the limitations of traditional media. This has the potential to enhance our enjoyment of entertainment and foster greater empathy and understanding between individuals from different walks of life.

Artificial intelligence, on the other hand, has the potential to breathe life into these virtual personalities, making them more authentic and responsive to our needs and desires. As AI continues to advance, we may find ourselves forming connections with virtual beings that are indistinguishable from those we share with real people. This raises important questions about the nature of human connection and the role that technology should play in our lives.

As we move forward into this brave new world, we must consider the ethical implications of these developments. While the potential benefits of enhanced parasocial interaction are undeniable, we must also be mindful of the potential pitfalls. As virtual personalities become more lifelike and our connections with them deepen, we must ensure that we do not lose sight of the importance of genuine human connection. It is crucial that we strike a balance between embracing technological advancements and preserving the essence of what makes us human.

The future of parasocial interaction is undoubtedly exciting, with virtual reality and artificial intelligence poised to reshape the media and entertainment landscape in ways we can only begin to imagine. As we continue to explore the possibilities that these technologies present, it is important that we do so with a sense of curiosity, responsibility, and humility. By acknowledging the potential challenges and working together to address them, we can ensure that the future of parasocial interaction is one that enriches our lives and brings us closer together rather than driving us further apart.

In conclusion, the digital age presents us with a unique opportunity to redefine the nature of parasocial interaction and forge new connections that transcend the boundaries of traditional media. As we navigate these uncharted waters, let us do so with a spirit of exploration, collaboration, and mindfulness, so that we may harness the power of virtual reality and artificial intelligence to create a brighter, more connected future for us all.

Chapter Summary

- The integration of virtual reality (VR) and artificial intelligence (AI) in the realm of parasocial interaction has the potential to revolutionize the way we form connections with virtual personalities and consume media content.
- VR technology allows for immersive experiences that can enhance the depth and intensity of parasocial relationships, while AI-driven characters can engage with users in a more natural and human-like manner.
- The development of emotional intelligence in AI systems can lead to stronger emotional bonds between users and virtual personalities, blurring the line between reality and the virtual world.
- Ethical considerations surrounding the future of parasocial interaction include the authenticity of virtual relationships, privacy and data security, and the potential for social isolation and erosion of human connection.
- The future of media and entertainment will likely see personalized and interactive content, the rise of virtual influencers and celebrities, and the need for regulation and ethical guidelines to ensure healthy parasocial relationships.
- Potential challenges in parasocial interaction include the risk of users becoming overly attached to virtual personalities, the creation of echo chambers, and the need to balance technological advancements with a genuine human connection.

- Navigating the uncharted waters of parasocial interaction in the digital age requires curiosity, responsibility, and humility, as well as a focus on striking a balance between embracing technological advancements and preserving the essence of human connection.
- The future of parasocial interaction holds exciting possibilities, and by addressing potential challenges and working together, we can harness the power of VR and AI to create a more connected and empathetic world.

10

STRATEGIES FOR MANAGING AND UNDERSTANDING PARASOCIAL RELATIONSHIPS

An image depicting the emotional bond between a person and their favorite fictional character, showcasing the psychological foundation of their parasocial relationship.

In today's hyper-connected world, the lines between reality and fantasy have become increasingly blurred. With the rise of social media, streaming platforms, and other digital technologies, we are now more connected than ever to the lives of celebrities, influencers, and even fictional characters. This has led to the emergence of a unique and fascinating phenomenon known as parasocial relationships. These one-sided connections, where an individual feels a strong emotional bond with a public figure or character, have become integral to our modern media landscape. As such, it is crucial for us to address and understand the complexities of these relationships, as they can significantly impact our emotional well-being, self-esteem, and overall mental health.

Parasocial relationships are not a new concept; they have existed since the advent of mass media. However, the rapid evolution of technology and the proliferation of social media platforms have amplified their prevalence and intensity. This makes it all the more important for us to delve deeper into the intricacies of these connections and develop strategies for managing and understanding them.

In this chapter, we will explore the various characteristics and types of parasocial interactions. We will also discuss techniques for establishing healthy boundaries in parasocial interactions. By doing so, we aim to provide readers with a comprehensive understanding of parasocial relationships and the tools necessary to navigate them in a balanced and mindful manner.

As we embark on this journey, it is essential to recognize that parasocial relationships are not inherently negative. In fact, they can serve as a source of comfort, inspiration, and even personal growth for many individuals. However, when these relationships become excessive or unhealthy, they can lead to negative consequences, such as feelings of isolation, low self-esteem, and even addiction-like behaviors. By addressing the importance of parasocial relationships and providing strategies for managing them, we hope to empower readers to embrace the future of parasocial interactions with awareness, balance, and a sense of personal agency.

Identifying the Characteristics and Types of Parasocial Interactions

Parasocial interactions are a unique and complex phenomenon that has gained increasing attention in recent years. As we continue to immerse ourselves in the world of media and technology, it is essential to understand the various characteristics and types of parasocial interactions to better manage and navigate these relationships. In this section, we will delve into the different aspects of parasocial interactions, shedding light on their nature and the ways in which they manifest in our lives.

Parasocial interactions share some common characteristics that set them apart from traditional interpersonal relationships. These characteristics include:

- **One-sidedness:** Parasocial relationships are typically one-sided, with the individual (or "fan") investing time, energy, and emotions into the relationship, while the media figure (e.g., celebrity, influencer, or fictional character) remains unaware of the individual's existence.
- **Illusion of intimacy:** Parasocial interactions often create a sense of intimacy and closeness between the individual and the media figure, despite the lack of direct communication or personal knowledge of each other.
- **Persistence:** These relationships can be long-lasting and resistant to change, with individuals maintaining their emotional attachment to the media figure even in the face of new information or changes in the media figure's public persona.
- **Emotional investment:** Individuals often experience strong emotions in response to the media figure's actions or perceived feelings, such as joy, sadness, anger, or empathy.

Parasocial interactions can be categorized into various types, depending on the nature of the relationship and the media figure involved. Some common types include:

- **Celebrity-based interactions:** These parasocial relationships involve an individual's emotional attachment to a famous person, such as an actor, musician, or athlete. The individual may follow the celebrity's career, personal life, and social media presence, experiencing a sense of connection and familiarity with the celebrity.
- **Influencer-based interactions:** With the rise of social media platforms, parasocial relationships with influencers have become increasingly common. Influencers are individuals who have gained a significant following on social media due to their expertise, lifestyle, or personal brand. Followers may feel a sense of connection and trust with the influencer, often seeking their advice or opinions on various topics.
- **Fictional character-based interactions:** Parasocial relationships can also form with fictional characters from books, movies, or television shows. Individuals may become emotionally invested in the character's story, experiencing a sense of connection and empathy for the character's experiences and emotions.
- **Online community-based interactions:** In some cases, parasocial interactions may extend beyond a single media figure to encompass an entire online community or fandom. This type of parasocial relationship involves a shared interest in a particular media figure or fictional universe, with individuals forming connections and emotional attachments to both the media figure and fellow fans within the community.

By understanding the characteristics and types of parasocial interactions, we can better recognize and manage these relationships in our lives. In the following sections, we will explore the psychological and emotional impact of parasocial relationships and discuss strategies for establishing healthy boundaries in these interactions.

Techniques for Establishing Healthy Boundaries in Parasocial Interactions

As parasocial interactions become increasingly prevalent in our lives, it is essential to establish healthy boundaries to ensure that these relationships do not negatively impact our well-being. In this section, we will discuss various techniques that can help individuals maintain a balanced approach to parasocial relationships, allowing them to enjoy the benefits while minimizing potential harm.

The first step in establishing healthy boundaries is to develop self-awareness and engage in regular reflection. By understanding our motivations and emotional responses to parasocial interactions, we can better recognize when these relationships are becoming unhealthy or detrimental to our well-being. It is essential to ask ourselves questions such as:

- What am I seeking from this parasocial relationship?
- How does this interaction make me feel?
- Am I investing too much time and emotional energy into this relationship?

By reflecting on these questions, we can gain insight into our parasocial interactions and make informed decisions about how to manage them effectively.

Another essential technique for establishing healthy boundaries is managing the time spent engaging in parasocial interactions. It is crucial to strike a balance between the time dedicated to parasocial relationships and other aspects of our lives, such as work, hobbies, and personal relationships. Setting limits on the amount of time spent consuming content or engaging with public figures can help prevent parasocial relationships from becoming all-consuming.

To maintain a healthy balance in our lives, it is essential to diversify our interests and relationships. Focusing solely on one parasocial relationship can lead to an unhealthy obsession and negatively impact our well-being. By engaging in various activities and maintaining a range of personal relationships, we can ensure that our lives are not solely defined by our parasocial interactions.

In addition to managing the time spent on parasocial interactions, it is crucial to establish emotional boundaries. This involves recognizing that the public figures we engage with may not reciprocate our feelings or be aware of our existence. It is essential to remind ourselves that parasocial relationships are one-sided and should not replace genuine personal connections.

If we struggle to establish healthy boundaries in our parasocial interactions, it may be helpful to seek support from friends, family, or a mental health professional. Sharing our experiences and discussing our feelings can provide valuable insights and help us develop strategies for managing our parasocial relationships in a healthy manner.

In conclusion, establishing healthy boundaries in parasocial interactions is crucial for maintaining our well-being and ensuring that these relationships do not negatively impact our lives. By developing self-awareness, managing our time, diversifying our interests, setting emotional boundaries, and seeking support when needed, we can enjoy the benefits of parasocial relationships while minimizing potential harm.

Embracing the Future of Parasocial Relationships with Awareness and Balance

As we have explored throughout this chapter, parasocial relationships are an inevitable part of our modern lives. With the ever-evolving media and technology landscape, these one-sided connections with public figures, celebrities, and even fictional characters will continue to grow and evolve. However, it is crucial for individuals to approach these relationships with awareness and balance to ensure that they do not negatively impact their mental and emotional well-being.

First and foremost, it is essential to recognize the difference between parasocial relationships and real-life connections. While parasocial interactions can provide comfort, entertainment, and inspiration, they should not replace the genuine bonds we form with friends, family, and loved ones. Maintaining a healthy balance between the two is key to ensuring that our emotional needs are met in a meaningful and fulfilling way.

Moreover, it is important to be aware of the potential pitfalls of parasocial relationships, such as excessive emotional investment, unrealistic expectations, and the blurring of boundaries between the public and private lives of celebrities. By understanding these risks, individuals can take proactive steps to manage their parasocial interactions in a healthy and responsible manner.

One way to achieve this balance is by implementing the techniques discussed in this chapter, such as establishing clear boundaries, engaging in self-reflection, and seeking support from others when needed. By doing so, individuals can enjoy the benefits of parasocial relationships while minimizing their potential negative effects.

Furthermore, as media and technology continue to shape the way we interact with public figures, it is crucial for both consumers and creators to be mindful of the role they play in fostering parasocial dynamics. This includes promoting responsible consumption of media, encouraging open dialogue about the impact of parasocial relationships, and advocating for ethical practices within the entertainment industry.

In conclusion, the future of parasocial relationships is undoubtedly complex and ever-changing. However, by embracing this phenomenon with awareness and balance, we can navigate the world of one-sided connections in a way that enriches our lives without compromising our emotional well-being. As we continue to explore the intricacies of parasocial interactions, we hope that this chapter has provided valuable insights and strategies for managing and understanding these unique relationships.

Chapter Summary

- Parasocial relationships are one-sided emotional connections with public figures, celebrities, or fictional characters, which have become increasingly prevalent due to advancements in media and technology.

- These relationships can have both positive and negative psychological and emotional impacts on individuals, providing comfort and inspiration but also potentially leading to isolation, low self-esteem, and addiction-like behaviors.
- Identifying the characteristics and types of parasocial interactions, such as one-sidedness, illusion of intimacy, persistence, and emotional investment, can help individuals better understand and manage these relationships.
- Establishing healthy boundaries in parasocial interactions is crucial for maintaining emotional well-being. Techniques for doing so include self-awareness and reflection, time management, diversifying interests and relationships, establishing emotional boundaries, and seeking support when needed.
- Media and technology have significantly shaped parasocial dynamics, making them more accessible and immersive than ever before. This includes the evolution of media, the role of technology in facilitating parasocial interactions, and the impact of media and technology on parasocial relationship formation.
- The constant exposure to curated content and social media algorithms can reinforce parasocial bonds, creating a feedback loop that strengthens parasocial attachments.
- It is essential for both consumers and creators to be mindful of their role in fostering parasocial dynamics, promoting responsible media consumption, open dialogue about the impact of parasocial relationships, and advocating for ethical practices within the entertainment industry.
- Embracing the future of parasocial relationships with awareness and balance can help individuals navigate these one-sided connections in a way that enriches their lives without compromising their emotional well-being.

THE COMPLEX WORLD OF PARASOCIAL INTERACTION

As we reach the final chapter of this enlightening exploration into the world of parasocial interaction, it is essential to pause and reflect on the journey we have taken together. Throughout this book, we have delved deep into the complexities of parasocial relationships, examining their origins, evolution, and impact on our lives. We have navigated through the psychological implications of these one-sided connections, and we have analyzed the role they play in modern society. Now, as we prepare to conclude our investigation, let us take a moment to appreciate the insights we have gained and the understanding we have achieved.

From the early days of television to the ever-evolving landscape of social media, parasocial interaction has grown and transformed alongside the media platforms that facilitate it. As we have discovered, these interactions are not merely a byproduct of our media consumption; they are an integral part of the human experience, reflecting our innate desire for connection and belonging. By examining the history and development of parasocial relationships, we have gained a deeper understanding of how these connections have become so pervasive in our lives.

Throughout our exploration, we have also delved into the psychological underpinnings of parasocial interaction, shedding light on the emotional and cognitive processes that drive our attachment to media figures. We have seen how these relationships can provide comfort, companionship, and even a sense of identity for those who engage in them. At the same time, we have also acknowledged the potential pitfalls and negative consequences that can arise from excessive or unhealthy parasocial involvement.

In examining the role of parasocial interaction in modern society, we have recognized its impact on various aspects of our lives, from our personal relationships to our consumption habits and even our political beliefs. We have also discussed the ethical considerations that arise from the power dynamics inherent in these one-sided connections and the responsibility of media creators and consumers to foster healthy and respectful parasocial interactions.

As we look to the future, we have considered the potential directions and research opportunities that lie ahead in the field of parasocial interaction. From the development of new media platforms to the ever-changing nature of celebrity and fame, the landscape of parasocial relationships will undoubtedly continue to evolve. By staying informed and engaged in this area of study, we can better understand and navigate the complex world of parasocial interaction.

In this concluding chapter, we will synthesize the key insights and findings from our journey, highlighting the multifaceted nature of parasocial interaction and its impact on our lives. As we embrace the complexity of this phenomenon, we can appreciate the richness and nuance it brings to our understanding of human connection in the digital age.

The Evolution of Parasocial Interaction: From Television to Social Media

As we have explored throughout this book, parasocial interaction has come a long way since its inception as a concept in the 1950s. This section will delve into the fascinating journey of parasocial interaction, from its roots in television to its current prominence in social media.

In the early days of television, viewers formed one-sided relationships with their favorite characters and personalities. These parasocial relationships were limited by the technology of the time, as television was a passive medium that allowed for little interaction between the audience and the on-screen figures. Nevertheless, viewers felt connected to these characters and personalities, often experiencing a sense of companionship and emotional attachment.

Fast forward to the present day, and the landscape of parasocial interaction has changed dramatically. The advent of the internet and social media platforms has transformed the way we consume media and engage with public figures. Social media allows for a more interactive experience, blurring the lines between the audience and the personalities they follow. This increased interactivity has led to a more complex and nuanced form of parasocial interaction.

Today, social media platforms such as Instagram, Twitter, and YouTube have become breeding grounds for parasocial relationships. These platforms provide users with a seemingly endless stream of content, allowing them to feel more intimately connected to the lives of their favorite celebrities, influencers, and content creators. The constant updates and behind-the-scenes glimpses into the lives of these personalities create an illusion of intimacy, fostering a sense of connection and emotional investment in their lives.

Moreover, the interactive nature of social media allows users to engage with these personalities directly through comments, likes, and direct messages. This level of interaction creates a feedback loop, where the personalities may respond to their audience, further strengthening the parasocial bond. In some cases, this interaction can even lead to a sense of friendship or kinship between the audience and the personality, despite the fact that they have never met in person.

The evolution of parasocial interaction from television to social media has also led to a shift in the types of personalities that attract parasocial relationships. While television primarily featured actors and traditional celebrities, social media has given rise to a new breed of influencers and content creators who may not have achieved fame through traditional means. These individuals often share more personal aspects of their lives, further fueling the sense of intimacy and connection experienced by their audience.

In conclusion, the evolution of parasocial interaction from television to social media has resulted in a more complex and multifaceted phenomenon. The increased interactivity and intimacy offered by social media platforms have transformed the way we engage with public figures, leading to stronger and more nuanced parasocial relationships. As we continue to explore the world of parasocial interaction, it is crucial to understand and appreciate the complexities of this ever-evolving phenomenon.

The Psychological Implications of Parasocial Relationships

As we have explored throughout this book, parasocial interaction has become an integral part of our daily lives, especially with the advent of social media. In this section, we will delve into the psychological implications of parasocial relationships, examining both the positive and negative aspects of these one-sided connections.

On the positive side, parasocial relationships can provide individuals with a sense of companionship and belonging. For those who may feel isolated or lonely, engaging with their favorite celebrities or influencers can offer a sense of connection and comfort. Research has shown that parasocial relationships can help to alleviate feelings of loneliness, particularly for those who have difficulty forming traditional social bonds. Furthermore, these relationships can also serve as a source of inspiration and motivation, as individuals may look up to their favorite public figures and aspire to emulate their success.

In addition to providing companionship and inspiration, parasocial relationships can also contribute to an individual's self-esteem and self-worth. By associating with successful and admired public figures, individuals may feel a sense of pride and accomplishment, even if their connection is purely one-sided. This phenomenon, known as the "basking in reflected glory" effect, can lead to increased self-confidence and a more positive self-image.

However, it is essential to recognize that parasocial relationships also have potential negative psychological implications. One of the primary concerns is the potential for individuals to become overly reliant on these one-sided connections, neglecting their real-life relationships and social interactions. This overreliance can lead to isolation and loneliness, as individuals may struggle to form genuine connections with others.

Another concern is the potential for individuals to develop unrealistic expectations and idealized perceptions of their favorite public figures. As we have discussed in previous chapters, social media often presents a curated and highly edited version of reality, leading individuals to believe that their favorite celebrities and influencers lead perfect lives. This idealization can result in feelings of inadequacy and dissatisfaction with one's own life, as individuals may constantly compare themselves to the unattainable standards set by their parasocial connections.

Furthermore, the intensity of parasocial relationships can sometimes lead to unhealthy levels of obsession and even stalking behaviors. In extreme cases, individuals may become so consumed by their one-sided connections that they lose touch with reality and engage in dangerous or harmful behaviors.

In conclusion, the psychological implications of parasocial relationships are complex and multifaceted. While these one-sided connections can provide individuals with a sense of companionship, inspiration, and self-worth, they also have the potential to lead to feelings of isolation, dissatisfaction, and even obsession. As we continue to navigate the ever-evolving world of parasocial interaction, individuals must maintain a healthy balance between their online connections and real-life relationships, ensuring that they do not become overly reliant on the comfort and validation provided by their favorite public figures.

The Role of Parasocial Interaction in Modern Society

As we navigate the intricate web of human relationships in the 21st century, it is crucial to acknowledge the significant role that parasocial interaction plays in our modern society. With the rapid advancement of technology and the proliferation of social media platforms, parasocial relationships have become an integral part of our daily lives, shaping our perceptions, emotions, and behaviors in ways that were once unimaginable.

The rise of celebrities and influencers has undoubtedly contributed to the prevalence of parasocial interaction in contemporary culture. These public figures, who often amass millions of followers on social media platforms, have the power to shape societal norms and values through their actions, words, and even their mere presence. As a result, fans often develop strong emotional attachments to these individuals, leading to the formation of parasocial relationships that can significantly impact their self-esteem, identity, and decision-making processes.

The role of parasocial interaction in modern society extends beyond the realm of entertainment and into the sphere of mental health and well-being. For some individuals, engaging in parasocial relationships can provide a sense of comfort, companionship, and emotional support, particularly in times of loneliness or social isolation. However, it is essential to recognize that these relationships can also have negative consequences, such as fostering unrealistic expectations, promoting unhealthy comparisons, and exacerbating feelings of inadequacy or low self-worth.

In today's consumer-driven society, parasocial interaction has become a powerful marketing tool for businesses and brands. By capitalizing on the emotional connections that fans have with their favorite celebrities and influencers, companies can effectively promote their products and services, ultimately driving sales and increasing brand loyalty. This commercialization of parasocial interaction raises important ethical questions about the exploitation of human emotions for financial gain and the potential for manipulation and deception in advertising.

Parasocial interaction has also found its way into the political arena, with politicians and activists leveraging their public personas to connect with supporters and advance their agendas. This phenomenon has led to the rise of "celebrity politicians" and the increasing importance of charisma and personal appeal in political campaigns. Additionally, parasocial interaction has played a crucial role in the success of various social movements, as activists use social media platforms to forge emotional connections with their followers and inspire collective action.

In conclusion, the role of parasocial interaction in modern society is multifaceted and far-reaching, influencing our personal lives, mental health, consumer behavior, and even our political landscape. As we continue to explore the complexities of this phenomenon, it is essential to remain mindful of both its potential benefits and pitfalls, striving to foster healthy and balanced relationships in our increasingly interconnected world.

Future Directions and Research in Parasocial Interaction

As we have explored the fascinating world of parasocial interaction throughout this book, it is evident that this phenomenon is deeply ingrained in our modern society. With the rapid advancement of technology and the ever-evolving landscape of media, it is crucial for researchers and industry experts to continue investigating the implications and potential consequences of parasocial relationships. This section will delve into the future directions and research possibilities in parasocial interaction.

One of the most pressing areas for future research is the impact of parasocial interaction on mental health. As we have discussed in previous chapters, parasocial relationships can provide a sense of companionship and emotional support for individuals, but they can also lead to negative consequences such as addiction, social isolation, and unrealistic expectations. Further research should focus on identifying the factors that contribute to both positive and negative outcomes and developing interventions and strategies to mitigate potential harm.

Another promising avenue for future research is the exploration of cultural differences in parasocial interaction. As media becomes increasingly globalized, it is essential to understand how different cultures perceive and engage with parasocial relationships. This could involve examining the role of cultural values, beliefs, and norms in shaping parasocial interactions and investigating how media portrayals of celebrities and influencers vary across cultures. Such research could provide valuable insights into the ways in which parasocial interaction is influenced by cultural context and may help to inform more culturally sensitive media practices.

The rise of virtual reality (VR) and augmented reality (AR) technologies also presents exciting opportunities for future research in parasocial interaction. As these technologies become more sophisticated and immersive, they have the potential to create even more realistic and compelling parasocial experiences. Researchers should explore the ways in which VR and AR may enhance or alter parasocial relationships, as well as the potential ethical concerns and psychological implications associated with these immersive experiences.

Finally, it is essential for future research to consider the role of parasocial interaction in the context of broader societal issues, such as the increasing polarization and fragmentation of society. As individuals become more reliant on social media and other forms of digital communication, it is important to understand how parasocial relationships may contribute to or alleviate these issues. For example, researchers could investigate whether parasocial interaction can foster empathy and understanding between individuals from different backgrounds or whether it may exacerbate existing divisions and biases.

In conclusion, the complex world of parasocial interaction offers a wealth of opportunities for future research and exploration. By continuing to investigate the psychological, cultural, technological, and societal aspects of parasocial relationships, we can deepen our understanding of this fascinating phenomenon and work towards harnessing its potential for positive change. As we embrace the complexity of parasocial interaction, we can strive to create a more connected, empathetic, and inclusive world.

Embracing the Complexity of Parasocial Interaction

As we reach the end of our exploration into the fascinating world of parasocial interaction, it is important to take a moment to reflect on the complexity and nuance that this phenomenon presents. Throughout this book, we have delved into the various aspects of parasocial interaction, from its evolution in the age of television to its current manifestation in social media. We have also examined the psychological implications of parasocial relationships and the role they play in our modern society. In this final section, we will embrace the complexity of parasocial interaction and consider how it will continue to shape our lives in the future.

One of the key takeaways from our journey is the understanding that parasocial interaction is not a one-dimensional concept. It encompasses a wide range of experiences and emotions, from the simple enjoyment of a favorite television show to the deep emotional connections we form with social media influencers. As such, it is crucial to recognize that parasocial interaction is not inherently good or bad; rather, it is a complex and multifaceted aspect of the human experience that can have both positive and negative consequences.

In embracing the complexity of parasocial interaction, we must also acknowledge the responsibility that comes with it. As consumers of media, we have the power to shape the way parasocial relationships develop and influence our lives. By being mindful of the content we consume and the connections we form, we can ensure that our parasocial interactions are healthy and enriching experiences.

Moreover, as creators and influencers, it is essential to recognize the impact our content can have on the lives of our audience members. By fostering genuine connections and promoting positive messages, we can help to create a more supportive and empowering media landscape.

Looking to the future, it is clear that parasocial interaction will continue to evolve and adapt to the changing media landscape. As technology advances and new platforms emerge, we can expect to see new forms of parasocial relationships develop, bringing with them new challenges and opportunities. It is our responsibility as researchers, creators, and consumers to continue exploring this complex phenomenon in order to better understand its implications and harness its potential for positive change.

In conclusion, the world of parasocial interaction is a rich and intricate tapestry of human experience. By embracing its complexity and recognizing the power it holds, we can work together to create a more connected, empathetic, and understanding society. As we move forward, let us continue to explore the depths of parasocial interaction and celebrate the unique and diverse connections that it fosters.

INFOTAINMENT UNLEASHED

EXPLORING THE EVOLUTION, IMPACT, AND FUTURE OF INTEGRATED ENTERTAINMENT AND INFORMATION SYSTEMS

INTRODUCTION TO INFOTAINMENT

In the age of information, the world has witnessed a remarkable transformation in the way we consume and process content. The rapid evolution of technology and the ever-growing influence of social media have given birth to a new era - the era of infotainment. This exciting and dynamic landscape has blurred the lines between information and entertainment, creating a unique blend that captivates audiences and keeps them hooked.

Infotainment has become an integral part of our daily lives as we constantly seek to be informed and entertained simultaneously. From news channels to social media platforms, infotainment has permeated every aspect of our lives, shaping the way we perceive and interact with the world around us. It has redefined the boundaries of traditional media, paving the way for innovative formats that cater to the diverse needs and preferences of modern audiences.

As we stand at the dawn of the infotainment era, it is crucial to understand the forces that drive this phenomenon and the impact it has on our society. This book aims to provide a comprehensive overview of the world of infotainment, delving into its origins, its various forms, and its challenges and opportunities. By exploring the multifaceted landscape of infotainment, we hope to equip readers with the knowledge and insights necessary to navigate this ever-evolving domain.

So, let us embark on this fascinating journey together as we unravel the mysteries of infotainment and discover the myriad ways in which it is transforming our lives. Welcome to the dawn of the infotainment era – a world where information and entertainment collide, creating a captivating and immersive experience that is reshaping the very fabric of our society.

Tracing the Roots of Infotainment

The term "infotainment" is a portmanteau of the words "information" and "entertainment," and it aptly describes the phenomenon of combining these two elements to create content that is both informative and engaging. To truly understand the roots of infotainment, we must first delve into the history of information dissemination and entertainment in various forms of media.

The concept of infotainment can be traced back to the early days of print media when newspapers began incorporating human interest stories and comic strips alongside hard news to appeal to a wider audience. As radio emerged as a popular medium in the early 20th century, news broadcasts were often interspersed with music and other forms of entertainment to keep listeners engaged. The advent of television further expanded the possibilities for infotainment, as news programs began to include features, interviews, and other entertaining segments to maintain viewer interest.

However, the true explosion of infotainment as we know it today can be attributed to the rise of cable television and the internet. The 24-hour news cycle and the proliferation of niche channels catering to specific interests increased the demand for content that could inform and entertain. This gave birth to an array of infotainment programs, ranging from talk shows and documentaries to reality TV and investigative journalism.

The internet has further revolutionized the infotainment landscape, as it has provided a platform for content creators to reach a global audience with ease. Social media, blogs, podcasts, and video-sharing platforms like YouTube have given rise to a new generation of infotainment stars who blend humor, storytelling, and information to engage their audiences. These digital platforms have also allowed for greater interactivity and personalization, enabling users to tailor their infotainment experiences to their specific interests and preferences.

As we trace the roots of infotainment, it becomes evident that the desire for engaging, informative content has been a constant throughout the evolution of media. What has changed, however, is the way in which this content is delivered and consumed. The rise of digital platforms has democratized the infotainment landscape, allowing for greater diversity and creativity in the content that is produced and shared. As we continue to explore the world of infotainment, it is important to recognize and appreciate the rich history that has shaped this fascinating phenomenon.

Navigating the World of Infotainment

The primary purpose of this book is to serve as a comprehensive guide for readers who wish to delve into the fascinating world of infotainment. As an industry expert, I have observed the rapid evolution of this field and its growing impact on various aspects of our lives. This book aims to provide a thorough understanding of the concept of infotainment, its origins, and its current manifestations in different industries.

This book is designed to cater to a diverse audience, including media professionals, students, educators, and anyone with a keen interest in understanding the convergence of information and entertainment. The book will equip readers with the knowledge and tools necessary to critically analyze and appreciate the various forms of infotainment that surround us.

In this age of information overload, it is crucial to understand the role that infotainment plays in shaping our perceptions, opinions, and decisions. This book will help readers develop a discerning eye, enabling them to differentiate between valuable information and mere distractions. By understanding the mechanics of infotainment, readers will be better equipped to navigate the complex media landscape and make informed choices about the content they consume.

Furthermore, this book will provide insights into the creative and business aspects of the infotainment industry, offering valuable advice for those who aspire to build a career in this field. Drawing from my own experiences and expertise, I will share practical tips and strategies for success in the ever-evolving world of infotainment.

In essence, this book is a comprehensive resource that aims to educate, inform, and inspire readers to embrace the infotainment revolution and make the most of the opportunities it presents.

Insights from an Industry Expert

As an author and industry expert, I have had the privilege of witnessing the evolution of infotainment firsthand. From the early days of radio and television to the current digital age, the landscape of infotainment has transformed dramatically. This transformation has not only impacted the way we consume information and entertainment but has also shaped our society in ways we could never have imagined.

Throughout my career, I have had the opportunity to work with various media organizations, technology companies, and content creators. This experience has provided me with a unique vantage point to observe the trends and developments in the infotainment industry. I have seen the rise of new platforms, the convergence of media, and the emergence of innovative storytelling techniques. These insights have allowed me to develop a deep understanding of the challenges and opportunities that lie ahead in the world of infotainment.

One of the key insights I have gained over the years is the importance of striking the right balance between information and entertainment. Infotainment, at its core, is a fusion of these two elements. However, it is essential to ensure that the information presented is accurate, relevant, and reliable while still engaging and entertaining. This delicate balance is what sets successful infotainment apart from mere sensationalism or dry, uninteresting content.

Another crucial aspect of infotainment is the role of technology in shaping the industry. As an industry expert, I have seen how technological advancements have revolutionized the way we create, distribute, and consume infotainment. From the rise of social media and streaming platforms to the development of virtual and augmented reality, technology has opened up new avenues for infotainment to thrive. Content creators and media organizations need to stay ahead of the curve and adapt to these changes to remain relevant in this ever-evolving landscape.

Lastly, I believe that the future of infotainment lies in its ability to foster a sense of global connectedness. In today's world, where information is readily available at our fingertips, infotainment has the power to bridge cultural, geographical, and linguistic divides. By presenting information in an engaging and entertaining manner, infotainment can help us understand and appreciate the diverse perspectives and experiences of people from around the world.

As we delve deeper into the world of infotainment in this book, I hope to share my insights and experiences as an industry expert to help you navigate this fascinating landscape. Together, we will explore the multifaceted world of infotainment and embrace the revolution that is transforming the way we consume information and entertainment.

Embracing the Infotainment Revolution

As we reach the end of this introductory chapter, it is essential to recognize the profound impact that infotainment has had on our society. The fusion of information and entertainment has revolutionized the way we consume content, engage with the world around us, and even perceive reality. Embracing the infotainment revolution is not only about acknowledging its existence but also understanding its implications and harnessing its potential for the betterment of our lives.

The infotainment era has undoubtedly brought about a paradigm shift in the media landscape. Traditional forms of media have been forced to adapt and evolve to stay relevant in this ever-changing environment. As a result, we have witnessed the rise of new media platforms, innovative storytelling techniques, and a plethora of content that caters to diverse interests and preferences. This transformation has expanded the horizons of creativity and democratized access to information and entertainment for millions of people across the globe.

However, the infotainment revolution is not without its challenges. The blurring of lines between fact and fiction, the proliferation of sensationalism, and the potential for manipulation and misinformation are some of the pressing concerns that need to be addressed. As industry experts, content creators, and consumers, we must strive to strike a balance between the allure of entertainment and the responsibility of providing accurate and reliable information. This involves fostering media literacy, promoting ethical journalism, and cultivating a discerning audience that can navigate the complex world of infotainment with ease and confidence.

In the subsequent chapters of this book, we will delve deeper into the various facets of the infotainment landscape, exploring its history, evolution, and impact on different sectors such as news, education, and advertising. We will also examine the role of technology in shaping the future of infotainment and discuss the emerging trends that are set to redefine the way we consume and engage with content.

As an author and industry expert, my goal is to provide you with valuable insights, thought-provoking discussions, and practical guidance that will help you navigate the world of infotainment with ease and confidence. By understanding the nuances of this fascinating domain, we can collectively embrace the infotainment revolution and harness its potential to inform, educate, and entertain in a responsible and meaningful manner.

So, let us embark on this exciting journey together and explore the captivating realm of infotainment, where information and entertainment converge to create a dynamic, engaging, and enriching experience for all.

1

EVOLUTION OF INFOTAINMENT SYSTEMS

A vintage car cruising down a neon-lit street in the 1980s.

Infotainment, a term coined by combining 'information' and 'entertainment,' has become an integral part of our daily lives. From the moment we wake up to the time we go to bed, we are constantly surrounded by various forms of infotainment. Be it through our smartphones, televisions, or in-car systems, infotainment has revolutionized the way we consume information and entertainment. This chapter delves into the fascinating journey of infotainment systems, tracing their evolution from the early days of innovation to the cutting-edge technology we experience today.

The concept of infotainment is not new; it has existed for centuries in various forms. However, the rapid technological advancements over the past few decades have significantly transformed the landscape of infotainment systems. The emergence of digital technology, in particular, has played a pivotal role in shaping the modern infotainment systems we have come to know and rely on.

As we explore the evolution of infotainment systems, we will closely examine the early innovations and pioneers who laid the groundwork for this industry. We will examine the transformative impact of digital technology on infotainment systems and how it has enabled the seamless integration of these systems into our modern lives. Furthermore, we will discuss the future of infotainment, highlighting the trends and predictions that are expected to shape the industry in the coming years.

The ongoing evolution of infotainment systems is a testament to human ingenuity and our insatiable appetite for information and entertainment. As we continue to push the boundaries of technology, infotainment systems will undoubtedly play an even more significant role in our lives, making the world a more connected, informed, and entertaining place to live in.

The Emergence of Infotainment: Early Innovations and Pioneers

As we delve into the fascinating world of infotainment, it is essential to understand its roots and early beginnings. The term "infotainment" is a portmanteau of "information" and "entertainment," which aptly describes the purpose of these systems: to inform and entertain simultaneously. In this section, we will explore the emergence of infotainment, highlighting the early innovations and pioneers that paved the way for the sophisticated systems we enjoy today.

The concept of infotainment can be traced back to the early days of mass media when newspapers, radio, and television began to blend information with entertainment. One of the first examples of infotainment was the newspaper comic strip, which combined humor and storytelling with current events and social commentary. This innovative format allowed readers to engage with the news in a more enjoyable and accessible way.

Radio soon followed suit, with programs such as Orson Welles' "The War of the Worlds" broadcast in 1938, which blurred the lines between fiction and reality, creating a captivating and immersive experience for listeners. As television emerged as a dominant medium, shows like "The Twilight Zone" and "60 Minutes" continued to push the boundaries of infotainment, offering thought-provoking content that both informed and entertained viewers.

The 1980s and 1990s saw the rise of cable television and the proliferation of specialized channels, which further fueled the growth of infotainment. Channels like MTV, ESPN, and CNN combined entertainment and information in innovative ways, catering to niche audiences and interests. This era also witnessed the birth of the 24-hour news cycle, which placed a premium on engaging and attention-grabbing content to keep viewers tuned in.

One of the most significant early pioneers of infotainment was Ted Turner, who founded CNN in 1980. Turner's vision was to create a news network that would provide continuous coverage of global events, with a mix of hard news, analysis, and human-interest stories. This groundbreaking approach revolutionized the way people consumed news and set the stage for the modern infotainment landscape.

Another notable figure in the evolution of infotainment is Oprah Winfrey, whose eponymous talk show debuted in 1986. "The Oprah Winfrey Show" skillfully blended celebrity interviews, personal stories, and expert advice, creating a unique and compelling format that resonated with millions of viewers. Winfrey's ability to connect with her audience and tackle complex issues with empathy and insight made her a trailblazer in the world of infotainment.

In summary, the emergence of infotainment can be attributed to the innovative efforts of early pioneers who recognized the potential of combining information and entertainment to create engaging and accessible content. From newspaper comic strips to groundbreaking television programs and networks, these early innovations laid the foundation for the sophisticated infotainment systems that have become an integral part of our modern lives.

The Rise of Digital Technology: Transforming Infotainment Systems

The advent of digital technology has played a pivotal role in the transformation of infotainment systems, revolutionizing the way we consume information and entertainment. This section will delve into the key developments and breakthroughs that have shaped the modern infotainment landscape, highlighting the impact of digital technology on various aspects of our lives.

The shift from analog to digital technology has been a game-changer for infotainment systems. In the early days, infotainment relied heavily on traditional media formats such as print, radio, and television. However, the introduction of digital technology has opened up new avenues for content creation, distribution, and consumption. The emergence of the internet, in particular, has had a profound impact on the way we access and engage with infotainment.

One of the most significant developments in the digital era has been the proliferation of online platforms and streaming services. These platforms have enabled users to access a vast array of content at their fingertips, ranging from news articles and podcasts to movies and music. The convenience and accessibility offered by these services have led to a surge in demand for infotainment, with consumers now able to access their favorite content anytime, anywhere.

Another key development in the digital transformation of infotainment systems has been the rise of social media. Platforms such as Facebook, Twitter, and Instagram have not only changed the way we communicate with one another but have also become important sources of infotainment. Social media has made it possible for users to share and discover new content while also providing a platform for content creators to reach a wider audience. The real-time nature of social media has also made it an invaluable tool for staying up-to-date with the latest news and events.

The evolution of digital technology has also led to the emergence of new devices and gadgets that have further enhanced the infotainment experience. Smartphones and tablets, for instance, have become indispensable tools for accessing infotainment on the go. These devices have also paved the way for developing various apps and software that cater to different infotainment needs, from news aggregators and e-readers to video streaming and gaming apps.

In addition to portable devices, digital technology has also transformed infotainment systems in our homes and vehicles. Smart TVs and voice-activated assistants like Amazon's Alexa and Google Home have made it easier than ever to access and control infotainment content, while in-car infotainment systems have evolved to include features such as GPS navigation, hands-free calling, and smartphone integration.

In conclusion, the rise of digital technology has transformed infotainment systems, reshaping how we access and engage with content. The integration of digital technology into various aspects of our lives has made infotainment more accessible, convenient, and personalized than ever before. As digital technology continues to advance, it is likely that infotainment systems will continue to evolve, offering even more immersive and engaging experiences for users.

The Integration of Infotainment Systems into Modern Life

As infotainment systems have evolved, they have become an integral part of our daily lives, seamlessly blending into various aspects of modern society. This integration has been facilitated by advancements in digital technology, which have allowed infotainment systems to become more accessible, user-friendly, and versatile. This section will explore how infotainment systems have permeated various aspects of our lives, from our homes and workplaces to our vehicles and public spaces.

One of the most significant ways in which infotainment systems have integrated into modern life is through the proliferation of smartphones and mobile devices. These devices have become our constant companions, providing instant access to a wealth of information and entertainment options at our fingertips. From streaming music and videos to browsing social media and playing games, smartphones have revolutionized the way we consume and interact with infotainment content.

In addition to smartphones, infotainment systems have also become a staple in our homes through the widespread adoption of smart TVs and streaming devices. These technologies have transformed the way we watch television and movies, offering a more personalized and interactive viewing experience. With the ability to access a vast library of content on-demand, consumers can now easily discover and enjoy a diverse range of programming that caters to their unique interests and preferences.

The integration of infotainment systems into our vehicles has also significantly enhanced our driving experience. Modern cars are now equipped with sophisticated infotainment systems that offer a multitude of features, such as GPS navigation, hands-free calling, and voice-activated controls. These systems provide drivers and passengers with access to information and entertainment while on the road and contribute to increased safety and convenience.

Public spaces, too, have seen the integration of infotainment systems in the form of digital signage and interactive kiosks. These technologies provide valuable information and entertainment to people in various settings, such as shopping malls, airports, and train stations. By offering real-time updates on news, weather, and transportation schedules, as well as interactive maps and directories, digital signage and kiosks help to enhance the overall experience of navigating and utilizing public spaces.

The integration of infotainment systems into modern life has undoubtedly enriched our daily experiences and made information and entertainment more accessible than ever before. As technology continues to advance, we can expect to see even more innovative and immersive infotainment solutions that will further transform the way we live, work, and play.

The Future of Infotainment: Trends and Predictions

As we progress in the digital age, infotainment systems are poised to become even more advanced, versatile, and integrated into our daily lives. In this section, we will explore some of the most exciting trends and predictions for the future of infotainment, offering a glimpse into what we can expect in the coming years.

One of the most significant trends in the future of infotainment is the increasing convergence of various technologies. As devices become more interconnected, we can expect infotainment systems to seamlessly integrate with other aspects of our lives, such as home automation, wearable technology, and even our vehicles. This will lead to a more unified and personalized user experience, allowing individuals to easily access their preferred content and services across multiple platforms.

Another key trend in the future of infotainment is the rise of artificial intelligence (AI) and machine learning. These technologies have the potential to revolutionize the way we interact with infotainment systems, enabling them to become more intuitive, responsive, and tailored to individual preferences. For example, AI-powered voice assistants could become an integral part of infotainment systems, allowing users to control their devices through natural language commands and even engage in more complex conversations with their devices.

The increasing prevalence of augmented reality (AR) and virtual reality (VR) technologies also promises to have a significant impact on the future of infotainment. These immersive technologies have the potential to transform the way we consume content, offering new and engaging ways to experience entertainment, education, and communication. For instance, AR could be used to overlay digital information onto the physical world, while VR could transport users into entirely new environments and experiences.

In addition to these technological advancements, the future of infotainment will also be shaped by evolving consumer preferences and expectations. As audiences become more discerning and demanding, infotainment providers will need to adapt their offerings to cater to these changing tastes. This could include a greater emphasis on personalization, interactivity, and high-quality content, as well as a focus on addressing the needs of diverse and global audiences.

Finally, it is important to consider the potential ethical and societal implications of the future of infotainment. As these systems become more sophisticated and integrated into our lives, questions surrounding privacy, security, and the potential for manipulation will become increasingly relevant. It will be crucial for industry leaders, policymakers, and consumers alike to engage in ongoing discussions about the responsible development and use of infotainment technologies.

In conclusion, the future of infotainment is undoubtedly exciting, with numerous technological advancements and evolving consumer preferences set to shape the landscape in the coming years. As we continue to witness the ongoing evolution of infotainment systems, it is essential to remain mindful of the potential challenges and opportunities that lie ahead, ensuring that these innovations are harnessed for the benefit of all.

The Ongoing Evolution of Infotainment Systems

As we have explored throughout this chapter, the evolution of infotainment systems has been a fascinating journey, marked by continuous innovation and adaptation to the ever-changing landscape of technology and consumer preferences. From the early days of radio and television to the current era of digital technology and the internet, infotainment systems have come a long way in terms of their capabilities, reach, and impact on our daily lives.

The ongoing evolution of infotainment systems is driven by a myriad of factors, including advancements in technology, the increasing demand for personalized and immersive experiences, and the need to cater to a diverse and global audience. As we look ahead, it is clear that infotainment systems will continue to evolve, offering even more sophisticated and engaging content that seamlessly integrates into our lives.

One of the key trends shaping the future of infotainment is the increasing convergence of various media platforms, enabling users to access content across multiple devices and formats. This trend is already evident in the growing popularity of streaming services and the rise of multi-platform content providers. As technology continues to advance, we can expect to see even more seamless integration between different media platforms, allowing users to consume infotainment content in a truly personalized and convenient manner.

Another important trend is the growing importance of data and analytics in shaping the content and delivery of infotainment systems. As more and more data is generated by users and their devices, content providers will be able to leverage this information to create highly targeted and relevant experiences for their audiences. This will not only enhance the overall quality of infotainment content but also help providers to better understand and cater to the needs and preferences of their users.

Furthermore, the future of infotainment systems will likely be marked by an increasing emphasis on interactivity and user-generated content. As users become more accustomed to engaging with content in a participatory manner, infotainment providers will need to find innovative ways to incorporate user input and feedback into their offerings. This could take the form of interactive storytelling, collaborative content creation, or even the integration of gaming elements into traditional media formats.

Lastly, as the world becomes more interconnected and globalized, infotainment systems will need to adapt to cater to a diverse range of cultural and linguistic preferences. This will involve not only the creation of content that is relevant and appealing to different audiences but also the development of technology that enables seamless translation and localization of content across different languages and regions.

In conclusion, the ongoing evolution of infotainment systems is an exciting and dynamic process that promises to bring about even more engaging, immersive, and personalized experiences for users around the world. As technology continues to advance and our understanding of user preferences and behavior deepens, we can look forward to a future where infotainment systems are an integral and enriching part of our daily lives.

Chapter Summary

- Infotainment, a combination of information and entertainment, has evolved significantly over the years, with digital technology playing a pivotal role in shaping modern infotainment systems.
- Early innovations in infotainment can be traced back to newspaper comic strips, radio programs, and television shows that blended information and entertainment, with pioneers like Ted Turner and Oprah Winfrey leading the way.
- The rise of digital technology, including the internet, online platforms, and streaming services, has transformed the way we consume infotainment, making it more accessible and convenient.
- Social media platforms have become important sources of infotainment, allowing users to share and discover new content while providing a platform for content creators to reach wider audiences.
- Infotainment systems have integrated into various aspects of our lives, including smartphones, smart TVs, in-car systems,

and public spaces, enhancing our daily experiences and making information and entertainment more accessible.

- The future of infotainment will likely see increasing convergence of technologies, the rise of artificial intelligence and machine learning, and the growing prevalence of augmented and virtual reality, offering more immersive and engaging experiences.
- Evolving consumer preferences and expectations will shape the future of infotainment, with a greater emphasis on personalization, interactivity, and high-quality content catering to diverse and global audiences.
- As infotainment systems continue to evolve, it is crucial to consider the potential ethical and societal implications, such as privacy, security, and the potential for manipulation, and engage in ongoing discussions about responsible development and use of infotainment technologies.

2

COMPONENTS OF MODERN INFOTAINMENT SYSTEMS

A futuristic smart city, where various Internet of Things (IoT) devices seamlessly communicate with each other.

In today's fast-paced world, the automotive industry has evolved beyond merely providing transportation. Modern vehicles have become sophisticated machines that cater to the needs and desires of their occupants, offering a plethora of features and technologies aimed at enhancing the driving experience. One such innovation that has gained immense popularity in recent years is the infotainment system. As the name suggests, infotainment is a fusion of information and entertainment designed to keep drivers and passengers informed, entertained, and connected while on the road.

Infotainment systems have come a long way since their inception, transforming from basic radio and CD players to advanced, multi-functional systems that integrate various technologies and features. These systems have become an essential aspect of modern vehicles, with many consumers considering them a crucial factor when purchasing a new car. In this chapter, we will delve into the components of modern infotainment systems, exploring the role of connectivity, multimedia and entertainment features, navigation and location-based services, and the integration of voice control and artificial intelligence.

At the heart of every infotainment system is a central display, often a touchscreen, which serves as the primary interface for the user. This display allows drivers and passengers to access and control a wide range of functions, from audio and video playback to navigation and communication features. The user interface is designed to be intuitive and user-friendly, ensuring that even those who are not tech-savvy can easily navigate the system.

One of the most significant advancements in infotainment systems is the integration of smartphone connectivity. This feature allows users to seamlessly connect their smartphones to the vehicle's infotainment system, enabling them to access their favorite apps, make phone calls, send and receive messages, and even stream music and podcasts directly from their devices. This connectivity has revolutionized the way we interact with our vehicles, making it easier than ever to stay connected and entertained while on the road.

As we continue to explore the components of modern infotainment systems, it becomes evident that these systems are more than just a luxury or convenience feature. They have become integral to the driving experience, providing valuable information, entertainment, and connectivity options that enhance our time spent behind the wheel. In the following sections, we will delve deeper into the various aspects of infotainment systems, shedding light on the technologies and innovations that make them an essential component of today's vehicles.

The Role of Connectivity in Infotainment Systems

In today's fast-paced world, staying connected has become essential to our daily lives. As technology continues to evolve, so does the demand for seamless connectivity, even when we are on the move. This is where modern infotainment systems come into play. These advanced systems, found in most contemporary vehicles, offer many features that entertain, inform, and assist drivers and passengers alike. One of the most crucial components of these systems is connectivity, which serves as the backbone for a multitude of functions and features.

Connectivity is the lifeblood of modern infotainment systems, as it enables the exchange of data between the vehicle, its occupants, and the outside world. This constant flow of information allows for real-time updates, personalized content, and enhanced communication capabilities. In essence, connectivity transforms the vehicle into an extension of our digital lives, ensuring we remain connected to the people and information that matter most to us.

There are several ways in which infotainment systems can establish and maintain connectivity. These include the following:

- **Cellular Connectivity**: This is the most common form of connectivity in infotainment systems. By leveraging existing cellular networks, vehicles can access the internet, enabling features such as real-time traffic updates, streaming music, and hands-free calling.
- **Wi-Fi Connectivity**: Some infotainment systems also offer built-in Wi-Fi capabilities, allowing passengers to connect

their devices to the internet without relying on their cellular data plans. This can be particularly useful for long road trips, where passengers may want to stream movies or play online games.

- **Vehicle-to-Everything (V2X) Communication:** This emerging technology enables vehicles to communicate with other vehicles, infrastructure, and even pedestrians, paving the way for a more connected and safer driving experience. V2X communication can provide real-time information on traffic conditions and road hazards, and even facilitate cooperative driving maneuvers.

Connectivity enables a wide range of services and applications that enhance the driving experience. Some of the most popular connected services include the following:

- **Real-Time Traffic Information**: By accessing up-to-date traffic data, infotainment systems can provide drivers with the most efficient routes, helping to save time and reduce fuel consumption.
- **Streaming Media:** Connectivity allows passengers to access their favorite music, podcasts, and audiobooks on-demand, ensuring they always have something to keep them entertained during their journey.
- **Social Media Integration:** Some infotainment systems offer integration with popular social media platforms, allowing drivers and passengers to stay connected with friends and family, share their location, or even post updates about their trip.
- **Remote Vehicle Services:** Connectivity enables remote access to various vehicle functions, such as starting the engine, adjusting climate controls, or locking and unlocking doors, all from a smartphone app.

In addition to entertainment and communication, connectivity is vital in enhancing vehicle safety and convenience. Connected services such as emergency assistance, roadside assistance, and vehicle diagnostics can provide drivers with peace of mind, knowing that help is just a button press away. Furthermore, connectivity enables the integration of advanced driver assistance systems (ADAS), which use data from various sensors and cameras to assist drivers in avoiding accidents and navigating complex driving situations.

In conclusion, connectivity is an indispensable component of modern infotainment systems, providing drivers and passengers with a wealth of information, entertainment, and convenience features. As technology continues to advance, we can expect even greater levels of connectivity and integration, paving the way for a more connected, informed, and enjoyable driving experience.

Multimedia and Entertainment Features

In today's fast-paced world, the demand for instant access to information and entertainment has never been greater. Modern infotainment systems have risen to the challenge, offering a plethora of multimedia and entertainment features that cater to the diverse needs and preferences of users. This section delves into the various multimedia and entertainment features that have become integral components of modern infotainment systems, enhancing the overall user experience and transforming the way we consume content.

One of the most fundamental aspects of any infotainment system is its ability to provide users with a rich audio and video experience. Modern systems support a wide range of formats, allowing users to enjoy their favorite music, podcasts, and videos with ease. In addition, high-quality speakers and advanced audio processing technologies ensure that users have an immersive and enjoyable listening experience, whether they are streaming content from their smartphones or playing it directly from the infotainment system's built-in storage.

The rise of streaming services has revolutionized the way we consume multimedia content. Modern infotainment systems have embraced this trend by integrating popular streaming services such as

Spotify, Apple Music, and Netflix, allowing users to access their favorite content on the go. This seamless integration offers users a vast library of music, movies, and TV shows and ensures they can enjoy personalized recommendations and curated playlists, further enhancing their entertainment experience.

In an increasingly connected world, social media and messaging platforms play a crucial role in our daily lives. Modern infotainment systems recognize this and offer integration with popular platforms such as Facebook, Twitter, and WhatsApp. This allows users to stay connected with their friends and family, receive real-time updates, and even share their experiences on the go, all without having to reach for their smartphones.

As technology continues to evolve, infotainment systems have started to incorporate gaming and interactive content, offering users a more engaging and immersive experience. From simple puzzle games to advanced virtual reality experiences, modern systems are equipped with the necessary hardware and software to support a wide range of gaming options. This helps keep passengers entertained during long journeys and opens up new possibilities for interactive learning and exploration.

A key aspect of modern infotainment systems is their ability to offer a high degree of customization and personalization. Users can tailor their systems to suit their preferences, from choosing the layout of the home screen to selecting their favorite color schemes and themes. Furthermore, advanced algorithms and artificial intelligence can analyze user behavior and preferences to offer personalized content recommendations, ensuring that the infotainment system remains relevant and engaging for each individual user.

In conclusion, multimedia and entertainment features have become essential components of modern infotainment systems, offering users a diverse and engaging experience that caters to their unique needs and preferences. As technology continues to advance, we can expect to see even more innovative and immersive features being integrated into infotainment systems, further transforming the way we consume content and stay connected on the go.

Navigation and Location-Based Services

In the era of smart devices and advanced technology, navigation and location-based services have become integral to modern infotainment systems. These services not only provide convenience and safety to drivers and passengers but also enhance the overall driving experience. In this section, we will delve into the various aspects of navigation and location-based services, their importance in infotainment systems, and the innovations that are shaping their future.

The primary purpose of navigation systems in infotainment is to provide accurate and real-time directions to drivers, ensuring that they reach their destinations efficiently and safely. With the integration of location-based services, these systems can offer personalized and context-aware information, such as nearby points of interest, traffic updates, and weather forecasts. This combination of navigation and location-based services has become a vital component of infotainment systems, as it significantly contributes to a seamless and enjoyable driving experience.

The backbone of navigation and location-based services is the Global Positioning System (GPS), which uses satellite signals to determine a vehicle's precise location. Modern infotainment systems employ advanced GPS receivers and mapping technologies to provide accurate and up-to-date maps, turn-by-turn directions, and real-time traffic information. Some systems even offer 3D mapping and augmented reality features, enhancing the visual representation of the route and making it easier for drivers to navigate unfamiliar roads.

The rise of smartphones and mobile applications has led to a significant shift in the way people access navigation and location-based services. Many drivers now prefer to use their smartphones for navigation, as they offer a familiar interface and access to popular third-party applications like Google Maps and Waze. To cater to this preference, infotainment systems have started to incorporate smartphone integration through technologies like Apple CarPlay and Android Auto, allowing users to access their preferred navigation apps and services directly from the infotainment system's display.

Another crucial aspect of navigation and location-based services is their integration with Advanced Driver Assistance Systems (ADAS). These systems use data from various sensors and cameras to provide features like adaptive cruise control, lane departure warnings, and parking assistance. By combining this data with navigation information, infotainment systems can offer route guidance that considers real-time traffic conditions, road hazards, and other factors that may affect the driving experience.

As technology continues to evolve, navigation and location-based services in infotainment systems are expected to become even more advanced and personalized. Innovations like 5G connectivity, machine learning, and the Internet of Things (IoT) will enable these systems to offer real-time updates, personalized recommendations, and seamless integration with other connected devices. Furthermore, the development of autonomous vehicles will likely lead to a significant transformation in the way navigation and location-based services are designed and utilized, as drivers will no longer need to focus on the road and can instead engage with the infotainment system in new and exciting ways.

In conclusion, navigation and location-based services play a crucial role in modern infotainment systems, providing drivers with accurate directions, real-time information, and personalized experiences. As technology continues to advance, we can expect these services to become even more sophisticated, further enhancing the driving experience and shaping the future of infotainment systems.

Integration of Voice Control and Artificial Intelligence

The integration of voice control and artificial intelligence (AI) has revolutionized the way we interact with infotainment systems, making them more user-friendly, efficient, and personalized. This section will delve into the importance of voice control and AI in modern infotainment systems, exploring their benefits and the potential for future advancements.

Voice control has become an essential feature in infotainment systems, allowing users to operate various functions without the need for manual input. This hands-free approach enhances user experience and contributes to safer driving by minimizing distractions. Voice control technology has evolved significantly over the years, with systems now able to understand and process complex commands, accents, and even different languages.

The integration of AI into voice control systems has further enhanced their capabilities. AI-powered voice assistants, such as Apple's Siri, Amazon's Alexa, and Google Assistant, have become increasingly popular in infotainment systems. These intelligent assistants can process natural language, learn from user interactions, and provide personalized recommendations based on user preferences and habits.

One of the most significant benefits of AI-powered voice control is its ability to adapt and improve over time. As users interact with the system, the AI learns from their preferences, speech patterns, and habits, allowing it to provide more accurate and relevant responses. This personalized approach ensures that the infotainment system remains engaging and useful throughout the life of the vehicle.

In addition to voice control, AI plays a crucial role in other aspects of infotainment systems. For example, AI algorithms can analyze user preferences and habits to curate personalized playlists, suggest new content, or even adjust the vehicle's climate control settings based on the user's comfort preferences. AI can also be used to optimize navigation routes based on real-time traffic data, weather conditions, and user preferences, ensuring a more efficient and enjoyable driving experience.

Looking ahead, the integration of voice control and AI in infotainment systems is expected to continue evolving. Advancements in natural language processing and machine learning will enable voice assistants to understand and process even more complex commands, making the user experience seamless and intuitive. Furthermore, as AI technology becomes more sophisticated, we can expect to see even greater levels of personalization and predictive capabilities in infotainment systems.

In conclusion, the integration of voice control and artificial intelligence has significantly enhanced the capabilities and user experience of modern infotainment systems. By allowing for hands-free operation, personalized recommendations, and adaptive learning, these technologies have made infotainment systems more engaging and useful than ever before. As voice control and AI technology continue to advance, we can expect to see even more innovative and exciting developments in the world of infotainment systems.

Chapter Summary

- Modern infotainment systems have evolved from basic radio and CD players to advanced, multi-functional systems that integrate various technologies and features, enhancing the driving experience.
- Connectivity is a crucial component of infotainment systems, enabling the exchange of data between the vehicle, its occupants, and the outside world, and providing real-time updates, personalized content, and enhanced communication capabilities.
- Multimedia and entertainment features have become essential components of modern infotainment systems, offering users a diverse and engaging experience that caters to their unique needs and preferences.
- Navigation and location-based services play a crucial role in modern infotainment systems, providing drivers with accurate directions, real-time information, and personalized experiences.
- The integration of voice control and artificial intelligence (AI) has revolutionized the way we interact with infotainment systems, making them more user-friendly, efficient, and personalized.
- AI-powered voice assistants, such as Apple's Siri, Amazon's Alexa, and Google Assistant, have become increasingly

popular in infotainment systems, providing personalized recommendations based on user preferences and habits.

- As concerns about data privacy and security continue to grow, future infotainment systems will likely place a greater emphasis on protecting users' information and developing more secure communication protocols and features that allow users to control the data collected and shared by their infotainment system.

3

INFOTAINMENT IN AUTOMOTIVE INDUSTRY

A futuristic infotainment system seamlessly integrated into the dashboard of a sleek, self-driving car.

Infotainment has become an essential aspect of the modern automotive industry. With rapid technological advancements and the increasing demand for connectivity and convenience, infotainment systems have evolved from simple radio receivers to sophisticated, multi-functional platforms that integrate navigation, communication, and entertainment features. These systems have transformed the way drivers and passengers interact with their vehicles, enhancing the overall driving experience and providing a seamless connection between the car and the digital world.

The automotive infotainment market has grown exponentially in recent years, driven by consumer demand for more advanced and user-friendly systems. As a result, car manufacturers and technology companies have been investing heavily in research and development to create innovative solutions that cater to their customers' ever-changing needs and preferences. This has led to the emergence of a highly competitive landscape, with industry players striving to differentiate themselves by offering unique and cutting-edge features that enhance the overall in-car experience.

In this chapter, we will delve into the fascinating world of automotive infotainment, exploring its evolution and development, key features and technologies, and the impact it has on driver experience and safety. We will also discuss future trends and innovations that are set to shape the future of infotainment in the automotive industry, highlighting the growing significance of this market segment in the broader context of the automotive ecosystem.

As we embark on this journey, it is important to recognize that infotainment systems are not just about providing entertainment and convenience; they also play a crucial role in improving driver safety and promoting responsible driving behavior. By offering real-time information, alerts, and assistance, these systems can help drivers make better decisions on the road, ultimately contributing to a safer and more enjoyable driving experience for all.

Evolution and Development of Automotive Infotainment Systems

The automotive infotainment industry has come a long way since its inception. From the early days of simple AM/FM radio systems to the sophisticated, interconnected systems of today, the evolution of automotive infotainment has been driven by technological advancements and the ever-changing demands of consumers. This section will explore the key milestones and developments that have shaped the landscape of automotive infotainment systems.

The journey of automotive infotainment began in the 1930s with the introduction of the first car radio. This simple AM radio system, developed by Motorola, marked the beginning of in-car entertainment and paved the way for future innovations. As technology progressed, FM radio was introduced in the 1950s, offering better sound quality and a wider range of stations for drivers to enjoy.

The 1960s saw the introduction of the 8-track tape player, which allowed drivers to listen to their favorite music on the go. This was followed by the cassette tape player in the 1970s, which quickly became the preferred choice for in-car audio due to its compact size and ease of use. The 1980s brought about the first in-car CD players, offering even better sound quality and the ability to store more music on a single disc.

As the 1990s rolled in, the focus shifted from audio entertainment to a more comprehensive infotainment experience. The introduction of GPS navigation systems revolutionized the way drivers navigated, providing real-time directions and maps. This decade also saw the rise of in-car television and DVD players, allowing passengers to watch movies and TV shows during long journeys.

The new millennium brought about a significant shift in the automotive infotainment landscape, with the advent of digital technology and the rise of the internet. The integration of Bluetooth technology enabled hands-free calling and wireless music streaming, while USB ports allowed for seamless connectivity with personal devices. The introduction of satellite radio provided drivers with an even wider range of entertainment options, including news, sports, and weather updates.

In recent years, the development of advanced infotainment systems has been heavily influenced by the proliferation of smartphones and mobile applications. Modern infotainment systems now offer touch-screen interfaces, voice recognition, and app integration, allowing drivers to access their favorite music, navigation, and communication apps directly from their vehicle's dashboard. These systems also support Apple CarPlay and Android Auto, further enhancing the connectivity between smartphones and in-car infotainment.

As we look toward the future, the evolution of automotive infotainment systems is expected to continue rapidly. With the advent of autonomous vehicles and the increasing importance of connectivity, we can expect to see even more advanced features and technologies integrated into our cars. From augmented reality navigation to AI-powered voice assistants, the possibilities for automotive infotainment are virtually limitless.

In conclusion, the evolution and development of automotive infotainment systems have been marked by significant technological advancements and a constant drive to meet the ever-changing demands of consumers. As we move forward, it is clear that infotainment will continue to play a crucial role in shaping the future of the automotive industry, enhancing the driving experience and offering new levels of connectivity and entertainment.

Key Features and Technologies in Modern Infotainment Systems

In today's fast-paced world, the automotive industry has evolved to meet the ever-changing needs and preferences of consumers. One such development is the integration of infotainment systems in vehicles, which has transformed the driving experience by providing a seamless blend of entertainment, information, and connectivity. This section delves into the key features and technologies that make modern infotainment systems indispensable to the contemporary driving experience.

Gone are the days of fumbling with buttons and knobs to control your car's entertainment system. Today's infotainment systems are equipped with high-resolution touchscreen displays allowing drivers and passengers to access various features with a tap or swipe. These displays come in various sizes and offer intuitive interfaces, making it easy for users to navigate through menus, adjust settings, and access information.

Smartphone integration is a crucial aspect of modern infotainment systems. Technologies such as Apple CarPlay and Android Auto enable drivers to connect their smartphones to the vehicle's infotainment system, allowing them to access their favorite apps, make phone calls, send messages, and stream music, all through the car's touchscreen display or voice commands. This integration enhances the driving experience and promotes safety by minimizing distractions.

Voice recognition technology has come a long way, and its integration into infotainment systems has made it easier for drivers to control various features without taking their hands off the wheel. By simply speaking commands, drivers can adjust the volume, change radio stations, make phone calls, and even input navigation destinations. This hands-free control greatly contributes to a safer and more enjoyable driving experience.

Built-in navigation systems have become a standard feature in many modern infotainment systems. These systems offer turn-by-turn directions, real-time traffic updates, and points of interest, ensuring that drivers can reach their destinations efficiently and safely. Some advanced navigation systems also incorporate augmented reality, displaying directional arrows and other information directly onto the windshield, further enhancing the user experience.

Connectivity is at the heart of modern infotainment systems, with features such as Wi-Fi hotspots, Bluetooth, and USB ports allowing drivers and passengers to stay connected on the go. These features enable users to access the internet, stream music, and connect multiple devices simultaneously, ensuring that everyone in the vehicle can enjoy a personalized and connected driving experience.

As technology continues to advance, infotainment systems need to keep up with the latest features and improvements. Over-the-Air (OTA) updates allow manufacturers to remotely update a vehicle's infotainment system, ensuring that drivers always have access to the most up-to-date software and features. This enhances the user experience and reduces the need for time-consuming and costly visits to dealerships for software updates.

In conclusion, modern infotainment systems have come a long way from their humble beginnings, offering a plethora of features and technologies that have revolutionized the driving experience. From touch-screen displays and smartphone integration to voice recognition and advanced navigation systems, these innovations have made driving more enjoyable, convenient, and safe. As technology continues to evolve, we can expect even more exciting developments in the world of automotive infotainment.

Impact of Infotainment on Driver Experience and Safety

The integration of infotainment systems in the automotive industry has significantly transformed the driving experience for motorists worldwide. These systems have enhanced the entertainment and connectivity aspects of driving and played a crucial role in improving driver safety. In this section, we will delve into the impact of infotainment on driver experience and safety, highlighting the benefits and potential challenges posed by these systems.

One of the most notable benefits of infotainment systems is the increased convenience and accessibility they offer to drivers. Modern infotainment systems provide a centralized hub for various functions like navigation, communication, and multimedia playback. This consolidation of features allows drivers to access essential information and perform tasks without having to fiddle with multiple devices or controls, thereby reducing distractions and enhancing overall safety.

Moreover, infotainment systems have significantly improved the quality of in-car entertainment. With features such as smartphone integration, streaming services, and advanced audio systems, drivers and passengers can now enjoy a wide range of entertainment options during their journeys. This makes long drives more enjoyable and helps alleviate driver fatigue, which is a significant contributor to road accidents.

Another critical aspect of infotainment systems is their role in promoting safer driving habits. Features such as voice recognition and hands-free controls enable drivers to perform tasks without taking their hands off the wheel or their eyes off the road. Additionally, advanced driver assistance systems (ADAS) integrated into infotainment systems provide real-time feedback on driving performance, alerting drivers to potential hazards and promoting safer driving behaviors.

However, it is essential to acknowledge the potential challenges and risks associated with infotainment systems. One such concern is the possibility of information overload, as drivers may be bombarded with too much data and notifications, leading to distractions and compromised safety. To mitigate this issue, automotive manufacturers must design intuitive and user-friendly interfaces that prioritize essential information and minimize distractions.

Another concern is the potential for over-reliance on infotainment systems and ADAS features, which may lead to complacency and reduced driver vigilance. Drivers need to understand that these systems are meant to assist and not replace their judgment and responsibility on the road.

In conclusion, the integration of infotainment systems in the automotive industry has significantly impacted driver experience and safety. While these systems offer numerous benefits, such as increased convenience, enhanced entertainment, and improved driving habits, it is crucial to address potential challenges and risks to ensure that infotainment systems continue to contribute positively to driver safety. As technology evolves, it is essential for automotive manufacturers and drivers alike to strike a balance between leveraging the benefits of infotainment systems and maintaining the necessary vigilance and responsibility on the road.

Future Trends and Innovations in Automotive Infotainment

As we venture further into the 21st century, the automotive industry continues to evolve rapidly, with infotainment systems playing an increasingly significant role in the overall driving experience. In this section, we will explore some of the most exciting and innovative trends that are set to shape the future of automotive infotainment.

One of the most promising developments in automotive infotainment is the integration of AI-powered voice assistants. These intelligent systems are designed to understand and respond to natural language commands, allowing drivers to interact with their vehicles in a more intuitive and hands-free manner. This not only enhances the user experience but also helps to minimize distractions and improve overall safety. As AI technology continues to advance, we can expect voice assistants to become even more sophisticated, capable of understanding complex requests and providing personalized recommendations based on user preferences and driving habits.

Another exciting innovation in the realm of automotive infotainment is the use of augmented reality and heads-up displays. These technologies project relevant information, such as navigation instructions, vehicle speed, and safety alerts, directly onto the windshield, allowing drivers to access critical data without taking their eyes off the road. As AR technology continues to mature, we can expect to see even more immersive and contextually-aware displays that seamlessly integrate with the surrounding environment, providing drivers with a truly futuristic driving experience.

The future of automotive infotainment also lies in the realm of vehicle-to-everything (V2X) communication. This technology enables vehicles to communicate with other vehicles, infrastructure, and even pedestrians, providing real-time information on traffic conditions, road hazards, and other relevant data. By leveraging V2X communication, infotainment systems can offer drivers more accurate and timely information, ultimately enhancing safety and efficiency on the road. As more vehicles become equipped with V2X capabilities, we expect to see a more connected and intelligent transportation ecosystem emerge.

As infotainment systems become more advanced, they will also offer greater levels of personalization and customization. This means that drivers can tailor their infotainment experience to their specific preferences, from the interface layout to the types of content and services available. Additionally, infotainment systems will be able to learn from user behavior and adapt accordingly, providing a more seamless and enjoyable experience for drivers and passengers alike.

Finally, the future of automotive infotainment will see greater integration with smart homes and Internet of Things (IoT) devices. This will enable drivers to control various aspects of their home environment, such as lighting, temperature, and security systems, directly from their vehicle's infotainment system. This connectivity level will provide added convenience and help create a more unified and cohesive user experience across multiple devices and platforms.

In conclusion, the future of automotive infotainment is bright, with numerous innovations and trends poised to revolutionize the way we interact with our vehicles. As technology continues to advance, we can expect to see even more sophisticated and immersive infotainment systems that enhance the driving experience while prioritizing safety and efficiency.

The Growing Significance of Infotainment in the Automotive Industry

In conclusion, the role of infotainment in the automotive industry has grown exponentially over the past few decades, transforming the way drivers and passengers interact with their vehicles. As technology continues to advance, infotainment systems have become an integral part of the driving experience, offering a wide range of features and functionalities that cater to the diverse needs and preferences of consumers.

The evolution of automotive infotainment systems has been driven by the rapid development of cutting-edge technologies, such as touchscreens, voice recognition, and smartphone integration. These advancements have enhanced the user experience and contributed to the overall safety and convenience of driving. By providing real-time traffic updates, navigation assistance, and hands-free communication, modern infotainment systems have significantly reduced the risk of driver distraction and improved situational awareness on the road.

Moreover, the growing significance of infotainment in the automotive industry can be attributed to the increasing consumer demand for seamless connectivity and personalized entertainment options. As a result, automakers are constantly striving to innovate and integrate new features into their infotainment systems, such as augmented reality, artificial intelligence, and advanced driver assistance systems (ADAS). These innovations are expected to further revolutionize the way we interact with our vehicles, making the driving experience more enjoyable, efficient, and safe.

In addition to enhancing the driving experience, the rise of infotainment systems has also created new opportunities for businesses within the automotive industry. The integration of various technologies and services has led to the emergence of new revenue streams, such as subscription-based content and data-driven services. Furthermore, the increasing importance of infotainment has also fostered collaboration between automakers, technology companies, and content providers, paving the way for new partnerships and business models.

As we look toward the future, it is evident that infotainment will continue to play a crucial role in shaping the automotive industry. With the advent of autonomous vehicles and the growing trend towards shared mobility, the role of infotainment systems is expected to evolve even further, providing passengers with a fully immersive and personalized experience. As a result, automakers and technology providers must continue to invest in research and development to stay ahead of the curve and meet the ever-changing needs of consumers.

In summary, the growing significance of infotainment in the automotive industry is a testament to the transformative power of technology and its ability to enhance our lives in countless ways. As we continue to witness the rapid evolution of infotainment systems, it is clear that they will remain an essential component of the driving experience, shaping the future of mobility and redefining our relationship with our vehicles.

Chapter Summary

- Infotainment systems have evolved from simple radio receivers to sophisticated, multi-functional platforms that integrate navigation, communication, and entertainment features, transforming the way drivers and passengers interact with their vehicles.
- The automotive infotainment market has grown exponentially due to consumer demand for advanced and user-friendly systems, leading to a highly competitive landscape with industry players striving to differentiate themselves through unique and cutting-edge features.
- The evolution of automotive infotainment systems has been driven by technological advancements, such as touchscreens, voice recognition, and smartphone integration, which have enhanced the user experience and contributed to overall safety and convenience.
- Modern infotainment systems offer many features and technologies, including touchscreen displays, smartphone integration, voice recognition and control, navigation systems, connectivity features, and over-the-air (OTA) updates.
- Infotainment systems have significantly impacted driver experience and safety by providing increased convenience, enhanced entertainment, and promoting safer driving habits through features such as voice recognition and hands-free controls.
- Future trends and innovations in automotive infotainment include the integration of artificial intelligence (AI) and voice assistants, augmented reality (AR) and heads-up displays (HUDs), vehicle-to-everything (V2X)

communication, personalization and customization, and integration with smart home and IoT devices.

- The growing significance of infotainment in the automotive industry has created new opportunities for businesses, fostering collaboration between automakers, technology companies, and content providers, and paving the way for new partnerships and business models.
- As the automotive industry continues to evolve, infotainment systems will remain an essential component of the driving experience, shaping the future of mobility and redefining our relationship with our vehicles.

4

USER INTERFACE DESIGN FOR INFOTAINMENT SYSTEMS

An infotainment system UI that takes inspiration from the vibrant colors and geometric shapes of the 1980s retro-futuristic aesthetic.

In today's fast-paced, technology-driven world, infotainment systems have become an integral part of our daily lives. From in-car entertainment and navigation systems to smart TVs and home automation devices, these systems provide a seamless blend of information and entertainment, making our lives more convenient, enjoyable, and connected. At the heart of these infotainment systems lies the user interface (UI), which serves as the bridge between the user and the technology. A well-designed UI is essential for ensuring that users can easily access and interact with the wealth of features and functionalities offered by these systems.

User interface design for infotainment systems is a specialized field that combines elements of graphic design, human-computer interaction, and cognitive psychology to create intuitive, visually appealing, and user-friendly interfaces. The primary goal of UI design in infotainment systems is to provide users with a seamless and enjoyable experience, allowing them to easily navigate the system, access relevant information, and control various features and functionalities with minimal effort.

As infotainment systems continue to evolve and become more sophisticated, the importance of effective UI design cannot be overstated. With an ever-growing array of features and functionalities, users are faced with an increasing amount of information and options, making it crucial for designers to create interfaces that are easy to understand, navigate, and interact with. Furthermore, as infotainment systems become more deeply integrated into our daily lives, the need for accessible and inclusive design becomes paramount, ensuring that users of all ages, abilities, and backgrounds can fully benefit from these technologies.

In this chapter, we will delve into the world of user interface design for infotainment systems, exploring the principles and best practices that guide the creation of effective, engaging, and accessible interfaces. We will discuss the delicate balance between aesthetics and functionality, as well as the importance of designing for accessibility and inclusivity. Through case studies of successful infotainment UI designs, we will examine the various elements and techniques that contribute to a seamless and enjoyable user experience. Finally, we will look ahead to the future of user interface design in infotainment systems, considering the emerging trends and technologies that will shape the way we interact with these systems in the years to come.

Principles of Effective Infotainment UI Design

In the rapidly evolving world of infotainment systems, user interface (UI) design plays a crucial role in ensuring that users can access and interact with content in a seamless and enjoyable manner. As infotainment systems become more sophisticated, designers must strike a balance between creating visually appealing interfaces and maintaining ease of use. This section will explore the fundamental principles of effective infotainment UI design, which can help designers create interfaces that look great and provide a satisfying user experience.

Clarity and simplicity: The primary goal of any UI design is to facilitate user interaction with the system. To achieve this, designers must prioritize clarity and simplicity in their designs. This means using clear, concise language, intuitive navigation, and easily recognizable icons and buttons. Designers can create a more streamlined and user-friendly interface by minimizing visual clutter and focusing on essential elements.

Consistency: Consistency is a vital aspect of UI design, as it helps users build a mental model of how the system works. By maintaining a consistent layout, color scheme, typography, and iconography across all screens and menus, designers can make it easier for users to navigate the interface and understand how different elements relate to one another.

Feedback and responsiveness: Infotainment systems must provide users with immediate feedback to confirm that their actions have been acknowledged and executed. This can be achieved through visual cues, such as highlighting a selected button or displaying a progress bar, as well as auditory cues, like a click or beep sound. Responsiveness is also crucial, as users expect the system to react quickly to their inputs. Delays in response can lead to frustration and a negative user experience.

Flexibility and customization: As infotainment systems cater to diverse users with varying preferences and needs, it is essential to offer flexibility and customization options. This can include adjustable font sizes, color themes, and the ability to rearrange or hide certain elements on the screen. Designers can enhance user satisfaction and engagement by providing users with the tools to tailor the interface to their liking.

Anticipating user needs: A well-designed infotainment UI should anticipate the user's needs and provide relevant information and options at the right time. This can be achieved through the use of context-sensitive menus, predictive text input, and smart recommendations based on user behavior and preferences. By proactively addressing user needs, designers can create a more efficient and enjoyable user experience.

Minimizing cognitive load: Infotainment systems often involve multitasking, such as navigating a map while listening to music or making a phone call. To prevent users from becoming overwhelmed, designers must minimize the cognitive load by presenting information in a clear, organized manner and avoiding information overload. This can be achieved through the use of hierarchical structures, grouping related elements together, and providing visual cues to guide users through the interface.

In conclusion, effective infotainment UI design is a delicate balance of aesthetics, functionality, and usability. By adhering to these principles, designers can create visually appealing interfaces and provide a satisfying and efficient user experience. As infotainment systems continue to evolve, it is crucial for designers to stay up-to-date with the latest trends and technologies to ensure that their designs remain relevant and engaging for users.

Balancing Aesthetics and Functionality in Infotainment Interfaces

In the world of infotainment systems, the user interface (UI) plays a crucial role in delivering a seamless and enjoyable experience for the end user. A well-designed UI enhances the system's overall aesthetics and ensures users can easily access and interact with the content. Striking the right balance between aesthetics and functionality is a critical aspect of UI design in infotainment systems, and this section will delve into the various considerations and best practices that designers should keep in mind while working on such projects.

First and foremost, it is essential to understand that aesthetics and functionality are not mutually exclusive. A visually appealing interface can significantly improve the user experience by creating a positive first impression and fostering a sense of trust and credibility. However, it is equally important to ensure that the interface is easy to navigate and use, as even the most visually stunning design can fall short if users struggle to find the information or features they need.

To achieve the perfect balance between aesthetics and functionality, designers should focus on the following fundamental principles:

Clarity and simplicity: The primary goal of any infotainment UI should be to make it easy for users to access and consume content. This can be achieved by using clear and concise language, intuitive navigation, and a clean layout that minimizes visual clutter. Designers should avoid using overly complex graphics or animations that may distract users from the main content.

Consistency: Maintaining a consistent visual language throughout the interface is crucial for creating a cohesive and professional look. This includes using a uniform color scheme, typography, and iconography that align with the overall brand identity. Consistency also extends to the layout and placement of various UI elements, ensuring that users can quickly familiarize themselves with the interface and navigate it with ease.

Responsiveness and adaptability: Infotainment systems are often accessed through a variety of devices, such as smartphones, tablets, and in-car displays. Designers must ensure the UI is responsive and adaptable to different screen sizes and resolutions, providing an optimal viewing experience for all users. This may involve creating multiple layouts or using flexible design elements that can automatically adjust to the available screen space.

User feedback: Providing users with clear and immediate feedback is essential for creating an engaging and interactive experience. This can be achieved through subtle visual cues, such as highlighting buttons when they are pressed or using animations to indicate that an action has been performed successfully. Feedback can also be provided through sound or haptic feedback, depending on the device and user preferences.

Accessibility: Designing for inclusivity is a crucial aspect of UI design, ensuring that infotainment systems can be used by individuals with varying abilities and needs. This may involve incorporating features such as adjustable font sizes, high-contrast color schemes, and alternative input methods, such as voice commands or gesture controls.

In conclusion, balancing aesthetics and functionality in infotainment interfaces is a delicate process that requires careful consideration of various design principles and user needs. By focusing on clarity, consistency, responsiveness, user feedback, and accessibility, designers can create visually appealing and highly functional infotainment systems that cater to a diverse audience and provide an enjoyable user experience.

Designing for Accessibility and Inclusivity in Infotainment Systems

In today's diverse and interconnected world, it is more important than ever to design infotainment systems that cater to the needs of all users, regardless of their age, abilities, or cultural background. Accessibility and inclusivity should be at the forefront of every user interface (UI) designer's mind when creating infotainment systems. This section will explore the importance of designing for accessibility and inclusivity, as well as provide practical tips and guidelines for achieving these goals in infotainment UI design.

The importance of accessibility and inclusivity in infotainment systems cannot be overstated. As the global population continues to age and the prevalence of disabilities increases, it is crucial that infotainment systems are designed to be usable by everyone, including those with visual, auditory, cognitive, or physical impairments. Furthermore, as the world becomes more interconnected, infotainment systems must be designed to accommodate users from various cultural backgrounds and language proficiencies. By designing for accessibility and inclusivity, UI designers can ensure that their infotainment systems are enjoyed by the widest possible audience and that no one is excluded from the benefits of these systems.

To design infotainment systems that are both accessible and inclusive, UI designers should adhere to the following guidelines:

Follow established accessibility standards: There are several established accessibility standards and guidelines, such as the Web Content Accessibility Guidelines (WCAG) and the ADA Standards for Accessible Design. These guidelines provide a wealth of information on how to design UI elements that are accessible to users with various disabilities.

Use clear and simple language: To accommodate users with varying language proficiencies and cognitive abilities, it is essential to use clear and simple language in all text elements of the infotainment system. This includes labels, instructions, and error messages. Additionally, providing translations or localization options can further enhance the inclusivity of the system.

Ensure high contrast and legibility: Users with visual impairments may struggle to read text or discern UI elements if there is insufficient contrast between the text and background colors. Designers should ensure that all text and UI elements have a high contrast ratio and are easily legible.

Provide alternative input methods: Some users may have difficulty using traditional input methods, such as touchscreens or buttons, due to physical impairments. Designers should consider incorporating alternative input methods, such as voice control or gesture recognition, to accommodate these users.

Design for various screen sizes and resolutions: Infotainment systems are used on a wide range of devices, from smartphones to large in-car displays. Designers should ensure their UI is adaptable and functional on various screen sizes and resolutions.

Offer customizable settings: Allowing users to customize the appearance and functionality of the infotainment system can greatly enhance its accessibility and inclusivity. This may include options for adjusting text size, color schemes, or audio settings.

By following these guidelines and prioritizing accessibility and inclusivity in the design process, UI designers can create infotainment systems that are enjoyable and usable by all users, regardless of their abilities or background. This not only benefits the users themselves but also contributes to a more inclusive and connected world.

Case Studies: Successful Infotainment UI Designs

This section will explore real-world examples of successful infotainment user interface (UI) designs. These case studies will provide valuable insights into the strategies and techniques employed by industry leaders to create engaging, user-friendly, and accessible infotainment systems.

Case Study 1: Apple CarPlay

Apple CarPlay is a prime example of a successful infotainment UI design. Launched in 2014, CarPlay allows users to connect their iPhones to their car's infotainment system, providing a seamless and familiar interface for accessing music, navigation, and communication apps.

One of the key strengths of CarPlay's UI design is its simplicity. The interface features large, easy-to-read icons and a minimalistic design that closely resembles the iPhone's home screen. This familiarity allows users to navigate the system with ease, reducing the learning curve and minimizing driver distraction.

Additionally, CarPlay's UI design prioritizes accessibility and inclusivity. The system supports Siri voice commands, enabling users with visual impairments or limited dexterity to interact with the infotainment system without needing to touch the screen. Furthermore, CarPlay is available in multiple languages, ensuring a broad range of users can enjoy its features.

Case Study 2: Tesla Model S Infotainment System

Tesla's Model S infotainment system is another shining example of successful UI design in the automotive industry. The system features a massive 17-inch touchscreen display that serves as the central hub for controlling various vehicle functions, including climate control, navigation, and entertainment.

The Model S infotainment system's UI design is characterized by its clean, modern aesthetic and intuitive layout. The interface is divided into two main sections: the top half displays navigation and media information, while the bottom half provides access to vehicle controls and settings. This layout enables users to quickly locate and access the desired functions with minimal distraction.

Tesla's infotainment system also incorporates several accessibility features, such as voice commands and customizable font sizes. These design choices ensure that a diverse range of users can comfortably interact with the system.

Case Study 3: Spotify for Android Auto

Spotify's integration with Android Auto demonstrates how a popular music streaming service can successfully adapt its UI design for an infotainment platform. The Android Auto version of Spotify features a simplified interface with large, easy-to-tap icons and limited on-screen options.

This streamlined design allows users to quickly access their favorite playlists, browse through recommended songs, and control playback without getting overwhelmed by excessive visual clutter. Moreover, Spotify for Android Auto supports voice commands, enabling users to search for songs, artists, and albums without taking their hands off the wheel.

In conclusion, these case studies showcase the importance of thoughtful UI design in creating successful infotainment systems. By prioritizing simplicity, familiarity, accessibility, and inclusivity, industry leaders like Apple, Tesla, and Spotify have developed infotainment interfaces that are visually appealing, highly functional, and user-friendly. As technology continues to evolve, we can expect to see even more innovative and engaging UI designs in the infotainment space.

The Future of User Interface Design in Infotainment Systems

As we have explored throughout this chapter, user interface design plays a crucial role in the development and success of infotainment systems. From adhering to the principles of effective UI design to ensuring accessibility and inclusivity, designers must continually adapt and innovate to meet the ever-evolving needs and expectations of users. As we look toward the future of user interface design in infotainment systems, several key trends and advancements are expected to shape the industry and redefine the way we interact with these systems.

One of the most significant trends in the future of UI design for infotainment systems is the increasing integration of artificial intelligence (AI) and machine learning. These technologies have the potential to revolutionize the way users interact with infotainment systems by enabling more personalized, context-aware, and adaptive experiences. For example, AI-driven voice assistants can provide a more natural and intuitive way for users to access information and control various functions within the system. Additionally, machine learning algorithms can analyze user behavior and preferences to tailor the interface and content recommendations, ensuring a more relevant and engaging experience.

Another key development in the future of infotainment UI design is the rise of augmented reality (AR) and virtual reality (VR) technologies. These immersive technologies can offer new and exciting ways for users to interact with infotainment systems, blurring the lines between the digital and physical worlds. For instance, AR-based interfaces can overlay digital information onto the user's real-world environment, providing contextually relevant information and enhancing the overall experience. Similarly, VR can transport users to entirely new environments, offering unique and engaging infotainment experiences that were previously unimaginable.

The increasing importance of data privacy and security will also play a significant role in shaping the future of user interface design for infotainment systems. As these systems collect and store vast amounts of user data, designers must prioritize the protection of this information and ensure that users have control over their data. This will likely involve the development of more transparent and user-friendly privacy settings, as well as the incorporation of advanced security measures to safeguard user data from potential threats.

Lastly, as the world becomes more interconnected, the future of infotainment UI design will likely involve greater integration with other devices and systems. This could include seamless connectivity with smartphones, wearables, and smart home devices, allowing users to access and control their infotainment systems from various touchpoints. This increased connectivity will require designers to create consistent and intuitive user experiences across multiple platforms, ensuring that users can easily navigate and interact with their infotainment systems, regardless of the device they are using.

In conclusion, the future of user interface design in infotainment systems is poised for significant advancements and transformations. By embracing emerging technologies, prioritizing data privacy and security, and ensuring seamless integration with other devices, designers can create innovative and engaging infotainment experiences that cater to the diverse needs and preferences of users. As we move forward, designers need to remain adaptable and open to change as the world of infotainment continues to evolve at a rapid pace.

Chapter Summary

- User interface (UI) design plays a crucial role in infotainment systems, ensuring users can easily access and interact with the wealth of features and functionalities offered by these systems.
- Effective infotainment UI design involves balancing aesthetics and functionality, prioritizing clarity, simplicity, consistency, feedback, responsiveness, flexibility, and anticipating user needs.
- Designers should strive to create visually appealing interfaces that are also user-friendly, easy to navigate, and provide a satisfying user experience.
- Accessibility and inclusivity are essential aspects of UI design, ensuring that infotainment systems cater to users of all ages, abilities, and backgrounds.
- Following established accessibility standards, using clear language, ensuring high contrast and legibility, providing alternative input methods, and designing for various screen sizes and resolutions are key guidelines for creating accessible and inclusive infotainment systems.
- Successful infotainment UI designs, such as Apple CarPlay, Tesla Model S Infotainment System, and Spotify for Android Auto, prioritize simplicity, familiarity, accessibility, and inclusivity to create engaging and user-friendly experiences.
- The future of UI design in infotainment systems will likely involve the integration of artificial intelligence, augmented reality, virtual reality, increased data privacy and security, and seamless connectivity with other devices and systems.
- Designers must remain adaptable and open to change as the world of infotainment continues to evolve rapidly, embracing emerging technologies and prioritizing user needs and preferences.

5

CONNECTIVITY AND INTEGRATION IN INFOTAINMENT

A futuristic smart city, where various Internet of Things (IoT) devices seamlessly communicate with each other.

The world of infotainment has come a long way since its inception. From the early days of radio broadcasts and black-and-white television sets to the current era of streaming services, smart devices, and the Internet of Things (IoT), the way we consume information and entertainment has undergone a massive transformation. At the heart of this evolution lies the ever-growing importance of connectivity and integration in infotainment systems.

Connectivity, in the context of infotainment, refers to the ability of various devices, platforms, and services to communicate and share information with one another. Integration, on the other hand, is the seamless blending of these connected elements to create a unified and user-friendly experience. Together, these two aspects have revolutionized the way we access, consume, and interact with information and entertainment content.

The journey of connectivity in infotainment began with the introduction of wired connections, such as radio and television broadcasts, which allowed users to access a limited range of content through dedicated devices. As technology advanced, we witnessed the emergence of cable and satellite television, which expanded the variety of content available to viewers and enabled the integration of additional services, such as on-demand programming and interactive features.

The advent of the internet marked a significant turning point in the evolution of connectivity and integration in infotainment. With the World Wide Web came the possibility of accessing an almost limitless array of content from around the globe and the ability to share and interact with that content in real time. This development paved the way for the rise of streaming services, social media platforms, and other digital channels that have since become integral components of our infotainment landscape.

In recent years, the proliferation of smart devices and IoT technology has further accelerated the growth of connectivity and integration in infotainment. Today, we can access our favorite content and services through many devices, such as smartphones, tablets, smart TVs, and even voice-activated assistants. This has led to a more personalized and immersive infotainment experience, as users can now tailor their content consumption to their preferences and enjoy it across multiple platforms.

As we look ahead, the future of connectivity and integration in infotainment promises to be even more exciting and dynamic. With the impending rollout of 5G networks and the continued development of cutting-edge technologies, we can expect to see even greater levels of interconnectivity and seamless integration between devices, platforms, and services.

In this chapter, we will delve deeper into the various facets of connectivity and integration in infotainment, exploring the role of smart devices and IoT, the importance of cross-platform compatibility, the impact of 5G and future technologies, and the challenges and opportunities that lie ahead. By understanding these aspects, we can better appreciate the transformative power of connectivity and integration in shaping the future of infotainment.

The Role of Smart Devices and IoT in Infotainment Integration

The rapid proliferation of smart devices and the Internet of Things (IoT) has revolutionized the way we consume and interact with information and entertainment. In this section, we will explore the role of these technologies in infotainment integration and how they have transformed the landscape of this industry.

Smart devices like smartphones, tablets, and smart TVs have become integral to our daily lives. These devices provide us with communication and productivity tools and serve as gateways to a vast array of infotainment content. From streaming music and video to accessing news and social media, smart devices have made it easier than ever to stay informed and entertained on the go.

The integration of infotainment into smart devices has led to the development of various applications and platforms that cater to users' diverse interests and preferences. For instance, streaming services like Netflix, Spotify, and YouTube have become household names, offering a plethora of content that can be accessed anytime, anywhere. Furthermore, the rise of voice-activated virtual assistants, such as Amazon's Alexa, Google Assistant, and Apple's Siri, has enabled users to access infotainment content through simple voice commands, making the experience even more seamless and convenient.

The Internet of Things (IoT) refers to the network of interconnected devices and objects that communicate and exchange data with each other. In the context of infotainment, IoT plays a crucial role in connecting various devices and platforms, enabling a more integrated and immersive experience for users.

One of the most prominent examples of IoT in infotainment is the connected car. Modern vehicles are equipped with advanced infotainment systems that integrate navigation, music, and communication features, all accessible through a central interface. By connecting to the internet and other devices, such as smartphones and wearables, these systems can provide real-time traffic updates, personalized music playlists, and hands-free communication options, enhancing the overall driving experience.

Similarly, IoT has also made its way into our homes, with smart speakers, smart TVs, and other connected appliances working together to create a seamless infotainment ecosystem. Users can now control their entertainment systems, receive news updates, and even manage their home environment through a single, unified platform.

The combination of smart devices and IoT has created a powerful synergy that has propelled the infotainment industry to new heights. This fusion has enabled users to access and control their infotainment content across multiple devices, creating a seamless and personalized experience.

For instance, a user can start watching a movie on their smart TV, pause it, and then resume watching it on their tablet or smartphone without missing a beat. Similarly, they can listen to their favorite podcast on their morning commute and then continue listening on their smart speaker when they arrive home. This level of integration and convenience has become the new norm in the infotainment industry, and it is only expected to grow as technology continues to advance.

In conclusion, the role of smart devices and IoT in infotainment integration cannot be overstated. These technologies have transformed the way we consume and interact with information and entertainment, making it more accessible, personalized, and immersive than ever before. As we look to the future, it is clear that the synergy between smart devices and IoT will continue to shape the infotainment landscape, offering exciting new possibilities and experiences for users around the world.

Cross-Platform Compatibility: Bridging the Gap Between Devices and Services

As infotainment systems continue to evolve, one of the most significant challenges developers and consumers face is ensuring cross-platform compatibility. This means creating a seamless experience for users, regardless of the device or service they are using. In this section, we will explore the importance of cross-platform compatibility in infotainment, the steps being taken to bridge the gap between devices and services, and the benefits it brings to both users and the industry as a whole.

Cross-platform compatibility is essential in today's interconnected world, where users expect to access their favorite content and services on any device at any time. This expectation extends to infotainment systems, which must be able to communicate with a wide range of devices, from smartphones and tablets to wearables and smart home appliances. By ensuring compatibility across platforms, infotainment systems can offer users a consistent and enjoyable experience, regardless of the device they are using.

One of the key challenges in achieving cross-platform compatibility is the sheer number of devices and services available on the market. With each device and service utilizing its own unique software, hardware, and communication protocols, creating a seamless experience can be a daunting task. To address this issue, industry leaders have begun to develop standardized frameworks and protocols that allow for easier integration between devices and services.

For example, the Automotive Grade Linux (AGL) project is an open-source initiative that aims to create a unified platform for automotive infotainment systems. By providing a common framework for developers, AGL helps to streamline the process of integrating various devices and services, ultimately leading to a more cohesive user experience.

Another approach to cross-platform compatibility is the use of application programming interfaces (APIs) and software development kits (SDKs). These tools allow developers to create applications and services that can easily communicate with one another, regardless of the underlying platform. By utilizing APIs and SDKs, infotainment systems can offer a wide range of features and services, from navigation and voice recognition to streaming music and social media integration.

The benefits of cross-platform compatibility in infotainment are numerous. For users, it means a more seamless and enjoyable experience, as they can access their favorite content and services on any device without worrying about compatibility issues. For developers, it simplifies the process of creating new applications and services, as they can build on existing frameworks and protocols. And for the industry as a whole, it fosters innovation and collaboration, as companies can work together to create new and exciting infotainment experiences.

In conclusion, cross-platform compatibility is a crucial aspect of modern infotainment systems, as it allows for seamless integration between devices and services. By embracing standardized frameworks, APIs, and SDKs, the industry can overcome the challenges associated with compatibility and deliver a more cohesive and enjoyable experience for users. As technology continues to advance, we can expect to see even greater levels of connectivity and integration in infotainment, ultimately leading to a more interconnected and immersive experience for all.

The Impact of 5G and Future Technologies on Infotainment Connectivity

The advent of 5G technology has the potential to revolutionize the infotainment industry by providing lightning-fast connectivity and seamless integration of various devices and services. As we move towards a world where everything is connected, the impact of 5G and other emerging technologies on infotainment cannot be overstated. In this section, we will explore how these advancements are shaping the future of infotainment connectivity and what it means for consumers and businesses alike.

One of the most significant advantages of 5G technology is its ability to offer unparalleled speed and bandwidth compared to its predecessors. With data transfer rates up to 100 times faster than 4G, 5G enables the streaming of high-quality content, such as 4K and 8K videos, with virtually no buffering or lag. This improvement in speed and bandwidth allows for a more immersive and interactive infotainment experience, as users can now access and consume content in real time without disruptions.

The increased speed and bandwidth of 5G technology also contribute to an enhanced user experience in infotainment. With faster connectivity, users can access a wider range of content and services, such as augmented reality (AR) and virtual reality (VR) applications, which were previously limited by the constraints of 4G networks. Furthermore, 5G enables the seamless integration of multiple devices, allowing users to switch between their smartphones, tablets, and other smart devices without any loss in quality or performance.

Another significant impact of 5G on infotainment connectivity is the rise of edge computing. By processing data closer to the source, edge computing reduces latency and improves the overall performance of infotainment systems. This technology allows for real-time data analysis and decision-making, which is crucial for applications such as autonomous vehicles and smart city infrastructure. As a result, infotainment systems can now provide users with more accurate and up-to-date information, enhancing their overall experience.

The rapid advancements in 5G and other emerging technologies have paved the way for new business models and opportunities within the infotainment industry. For instance, the increased speed and bandwidth provided by 5G networks enable content creators to offer more immersive and interactive experiences, such as live events and virtual concerts. Additionally, businesses can now leverage the power of 5G to develop innovative solutions and services, such as personalized advertising and location-based content delivery, which were previously not feasible due to the limitations of 4G networks.

As the infotainment industry continues to evolve, it is crucial for businesses and consumers alike to stay informed about the latest developments in connectivity and integration technologies. By embracing the potential of 5G and other emerging technologies, we can unlock new possibilities and experiences in infotainment, ultimately transforming how we consume and interact with content.

In conclusion, the impact of 5G and future technologies on infotainment connectivity is profound and far-reaching. With faster speeds, enhanced user experiences, and new business opportunities on the horizon, the future of infotainment is undoubtedly exciting and full of potential. As we continue to push the boundaries of what is possible, the infotainment industry will undoubtedly continue to evolve and adapt to the ever-changing landscape of connectivity and integration.

Challenges and Opportunities in Ensuring Seamless Infotainment Integration

As the infotainment landscape continues to evolve, the need for seamless integration between various devices, platforms, and services becomes increasingly crucial. This integration is essential for providing users with a cohesive and enjoyable experience. However, achieving this level of integration is not without its challenges. In this section, we will explore some of the key obstacles faced by the industry in ensuring seamless infotainment integration, as well as the opportunities that arise from overcoming these challenges.

One of the primary challenges in achieving seamless infotainment integration is the sheer diversity of devices and platforms available in the market. From smartphones and tablets to smart TVs and gaming consoles, users have a wide range of options when it comes to accessing infotainment content. Furthermore, each device runs on different operating systems, such as Android, iOS, and Windows, which adds another layer of complexity to the integration process.

To overcome this challenge, developers and service providers must adopt a cross-platform approach, ensuring their applications and content are compatible with multiple devices and operating systems. This may involve using universal programming languages like HTML5 or developing dedicated apps for each platform. By doing so, they can provide users with a consistent experience across all devices, ultimately enhancing user satisfaction and engagement.

As infotainment systems become more connected and integrated, concerns surrounding data privacy and security inevitably arise. Users are often required to share personal information, such as their location, preferences, and usage habits, to access personalized content and services. If not properly protected, this data can be vulnerable to breaches and misuse.

To address these concerns, industry players must prioritize data privacy and security in their integration efforts. This can be achieved by implementing robust encryption technologies, secure data storage solutions, and strict privacy policies. By protecting users' data, service providers can build trust and encourage greater adoption of their infotainment offerings.

A lack of standardization and interoperability between devices, platforms, and services is another challenge that hinders seamless infotainment integration. Without common standards and protocols, it becomes difficult for different systems to communicate and share information effectively.

Industry stakeholders must collaborate to develop and adopt universal standards and protocols that facilitate seamless integration. This may involve the creation of industry-wide consortiums or the adoption of existing standards, such as those developed by the World Wide Web Consortium (W3C). By fostering greater standardization and interoperability, the industry can pave the way for more seamless and user-friendly infotainment experiences.

Despite the challenges, there are significant opportunities for those who can successfully achieve seamless infotainment integration. By providing users with a cohesive and enjoyable experience across multiple devices and platforms, service providers can increase user engagement, drive customer loyalty, and generate higher revenues.

Moreover, seamless integration can open up new avenues for innovation and collaboration within the industry. As devices and services become more interconnected, there is potential for the development of new business models, partnerships, and revenue streams. For example, the integration of infotainment services with smart home systems could enable users to control their home environment through their entertainment devices, creating new opportunities for both industries.

In conclusion, the future of connectivity and integration in infotainment is full of challenges and opportunities. By addressing the obstacles related to device diversity, data privacy, and standardization, industry players can unlock the full potential of seamless infotainment experiences and capitalize on the opportunities that lie ahead.

The Future of Connectivity and Integration in Infotainment

As we have explored throughout this chapter, the world of infotainment is rapidly evolving, driven by advancements in technology and the ever-growing demand for seamless connectivity and integration. The future of infotainment is undoubtedly exciting, as we can expect even more immersive and personalized experiences that cater to our individual preferences and needs.

One of the key factors shaping the future of connectivity and integration in infotainment is the continued development of smart devices and the Internet of Things (IoT). As these technologies become more sophisticated and widespread, we can anticipate a more interconnected world where infotainment systems are seamlessly integrated with our daily lives. This will enable us to access information, entertainment, and services in real time, regardless of our location or device.

Another major driving force in the future of infotainment is the advent of 5G and other emerging technologies. With faster and more reliable connections, we can expect a significant improvement in the quality and variety of infotainment experiences. This will enhance existing services and pave the way for new and innovative applications that were previously unimaginable.

Despite the numerous benefits and opportunities that come with enhanced connectivity and integration, there are also challenges that need to be addressed. Ensuring data privacy and security will be of paramount importance as infotainment systems become more interconnected and store increasing amounts of personal information. Additionally, the industry will need to tackle issues related to cross-platform compatibility and standardization to ensure a smooth and consistent user experience across different devices and services.

In conclusion, the future of connectivity and integration in infotainment is bright, with endless possibilities for innovation and growth. As technology continues to advance rapidly, we can look forward to a world where infotainment systems are an integral part of our daily lives, providing us with a seamless and personalized experience that keeps us informed, entertained, and connected. By addressing the challenges and embracing the opportunities that lie ahead, the infotainment industry can continue to thrive and shape the way we interact with the world around us.

Chapter Summary

- The evolution of connectivity and integration in infotainment has transformed the way we access, consume, and interact with information and entertainment content, from radio broadcasts to the Internet of Things (IoT).
- Smart devices and IoT play a crucial role in infotainment integration, enabling a more personalized and immersive experience for users by connecting various devices and platforms.
- Cross-platform compatibility is essential for a seamless infotainment experience, requiring developers to ensure their applications and content are compatible with multiple devices and operating systems.

- The advent of 5G technology promises to revolutionize the infotainment industry by providing lightning-fast connectivity, enabling more immersive and interactive experiences, and fostering the rise of edge computing.
- Ensuring data privacy and security is a significant challenge in infotainment integration, requiring robust encryption technologies, secure data storage solutions, and strict privacy policies.
- Standardization and interoperability between devices, platforms, and services are necessary for seamless infotainment integration, requiring industry stakeholders to collaborate and adopt universal standards and protocols.
- Seamless infotainment integration offers significant opportunities for increased user engagement, customer loyalty, and revenue generation, as well as fostering innovation and collaboration within the industry.
- The future of connectivity and integration in infotainment is bright, with endless possibilities for innovation and growth, as long as the industry addresses challenges related to data privacy, cross-platform compatibility, and standardization.

6

INFOTAINMENT AND THE INTERNET OF THINGS (IOT)

A futuristic smart city, where various Internet of Things (IoT) devices seamlessly communicate with each other.

In today's fast-paced world, the demand for information and entertainment is at an all-time high. People are constantly seeking new ways to stay informed, engaged, and entertained, leading to the rise of infotainment – a blend of information and entertainment. Infotainment is a rapidly growing industry that encompasses various forms of media, including television, radio, print, and digital platforms. It aims to provide audiences with a captivating experience that combines the informative aspects of news and educational content with the entertaining elements of movies, music, and games.

As technology continues to advance, the Internet of Things (IoT) has emerged as a game-changer in the world of infotainment. IoT refers to the interconnection of everyday objects and devices through the internet, allowing them to communicate and exchange data with one another. This widespread connectivity has the potential to revolutionize the way we consume and interact with infotainment, making it more personalized, immersive, and accessible than ever before.

The integration of infotainment and IoT has opened up a world of possibilities, transforming the way we experience entertainment and access information. In this chapter, we will explore the evolution of infotainment systems in the IoT era, examining how these technologies are converging to create a seamless and interconnected experience for users. We will also discuss the integration of infotainment with smart home and smart city technologies, highlighting the potential for even greater personalization and customization of our entertainment experiences.

As with any technological advancement, the fusion of infotainment and IoT also raises concerns about security and privacy. We will delve into these issues, discussing the potential risks and challenges associated with IoT-based infotainment systems and the measures to address them.

Finally, we will look ahead to the future of infotainment in an interconnected world, considering the exciting possibilities and potential pitfalls that lie ahead as we continue to embrace the power of IoT. So, sit back, relax, and join us on this fascinating journey into the world of infotainment and the Internet of Things.

Evolution of Infotainment Systems in the IoT Era

The dawn of the Internet of Things (IoT) has brought about a significant transformation in the world of infotainment. As the IoT era unfolds, infotainment systems have evolved from being standalone entertainment devices to becoming an integral part of a connected ecosystem. This evolution has enhanced the user experience and opened up new avenues for innovation and growth in the infotainment industry.

In the early days, infotainment systems were primarily focused on providing entertainment and information through traditional mediums such as radio, television, and print. However, with the advent of the internet and the proliferation of smartphones, the infotainment landscape witnessed a paradigm shift. Users now had access to a plethora of content and services at their fingertips, leading to the emergence of new infotainment platforms such as social media, streaming services, and online news portals.

The IoT era has further revolutionized the infotainment industry by enabling seamless integration of infotainment systems with various connected devices and platforms. This has led to the development of advanced infotainment systems that offer users a more immersive and personalized experience. Some key milestones in the evolution of infotainment systems in the IoT era include:

Connected Cars: Modern vehicles are now equipped with sophisticated infotainment systems that offer a wide range of services such as navigation, real-time traffic updates, voice control, and smartphone integration. These systems leverage IoT technology to communicate with other connected devices, enabling features such as remote diagnostics, vehicle tracking, and over-the-air software updates.

Smart TVs: The advent of smart TVs has transformed the traditional television viewing experience by offering users access to a vast array of online content and services. Smart TVs now come with built-in

IoT capabilities, allowing users to control various smart home devices and access personalized content recommendations based on their viewing habits.

Voice Assistants: Voice-controlled digital assistants such as Amazon's Alexa, Google Assistant, and Apple's Siri have become integral to the infotainment ecosystem. These AI-powered assistants leverage IoT technology to interact with various connected devices, enabling users to access information, control smart home appliances, and even make purchases using simple voice commands.

Wearable Technology: Wearable devices such as smartwatches and fitness trackers have added a new dimension to the infotainment experience by offering users access to personalized content and services on the go. These devices leverage IoT technology to collect and analyze user data, providing insights into their health, fitness, and lifestyle habits.

As the IoT era continues to unfold, the infotainment landscape is poised for further growth and innovation. The seamless integration of infotainment systems with various connected devices and platforms will not only enhance the user experience but will also create new opportunities for content creators, service providers, and technology companies to thrive in this interconnected world.

Integration of Infotainment with Smart Home and Smart City Technologies

Rapid technological advancements have led to the seamless integration of infotainment systems with smart home and smart city technologies. This integration has not only enhanced the user experience but has also transformed the way we consume information and entertainment. In this section, we will delve into the various ways infotainment systems are being integrated into smart home and smart city technologies, and how this integration is shaping the future of infotainment.

Smart homes are equipped with advanced automation systems that allow homeowners to control various aspects of their homes, such as lighting, temperature, and security, using a single device or application.

Infotainment systems have become an integral part of smart homes, providing users with a centralized platform to access and control their entertainment options.

- **Centralized Control:** With the integration of infotainment systems in smart homes, users can now control their entertainment devices, such as televisions, gaming consoles, and audio systems, from a single interface. This centralized control simplifies the user experience and allows for seamless interaction between different devices.
- **Voice-activated Assistance:** The incorporation of voice-activated assistants like Amazon's Alexa, Google Assistant, and Apple's Siri into infotainment systems has made it easier for users to access and control their entertainment options. Users can now simply use voice commands to play music, watch movies, or even get the latest news updates.
- **Personalized Recommendations:** Infotainment systems in smart homes can now analyze user preferences and habits to provide personalized content recommendations. This ensures that users are always presented with relevant and engaging content, enhancing their overall entertainment experience.

Smart cities are urban areas that utilize advanced technology and data analysis to improve the quality of life for their residents. Infotainment systems play a crucial role in smart cities, providing residents with real-time information and entertainment options that cater to their needs and preferences.

- **Public Information Displays:** Infotainment systems integrated into smart city infrastructure can provide residents with real-time information on various aspects of city life, such as weather updates, traffic conditions, and public transportation schedules. These displays can be found in public spaces like parks, bus stops, and train

stations, ensuring that residents have access to relevant information at all times.
- **Interactive Kiosks:** Smart cities are increasingly adopting interactive kiosks that provide residents with access to infotainment services. These kiosks can be used to access local news, events, and entertainment options, as well as provide directions and information about nearby attractions and amenities.
- **Connected Vehicles:** The integration of infotainment systems in connected vehicles has transformed the in-car entertainment experience. Drivers and passengers can now access real-time traffic updates, navigation services, and personalized entertainment options, making their journeys more enjoyable and efficient.

In conclusion, the integration of infotainment systems with smart home and smart city technologies has significantly enhanced the user experience, providing individuals with personalized and relevant content that caters to their needs and preferences. As technology continues to evolve, we can expect to see even more innovative ways in which infotainment systems are integrated into our daily lives, shaping the future of information and entertainment consumption.

Personalization and Customization of Infotainment Experiences through IoT

The Internet of Things (IoT) has revolutionized the way we interact with technology, allowing for the seamless integration of devices and systems in our daily lives. This interconnectedness has also profoundly impacted the world of infotainment, enabling users to enjoy highly personalized and customized experiences tailored to their preferences and needs. This section will explore how IoT has transformed the infotainment landscape, offering users a more immersive and engaging experience.

One of the most significant benefits of IoT in infotainment is the ability to collect and analyze vast amounts of user data. By gathering information on users' preferences, habits, and interests, IoT-enabled infotainment systems can create personalized content recommendations and experiences. For example, a smart TV connected to the IoT can analyze a user's viewing history and suggest movies or TV shows based on their preferences. Similarly, a connected car infotainment system can learn a driver's favorite routes, music preferences, and even preferred cabin temperature, ensuring a comfortable and enjoyable driving experience.

Another aspect of personalization and customization in IoT-based infotainment systems is the integration of voice assistants and natural language processing technologies. This allows users to interact with their devices using voice commands, making it easier to access content and control various aspects of the infotainment experience. For instance, a user can ask their smart speaker to play a specific song, provide news updates, or control smart home devices, all without lifting a finger. This level of convenience and personalization has become increasingly important as users seek more intuitive and user-friendly ways to interact with technology.

The IoT also enables infotainment systems to be more context-aware, adapting to the user's environment and situation. For example, a connected car infotainment system can detect when the vehicle is in heavy traffic and automatically switch to a more relaxing playlist or provide real-time traffic updates to help the driver navigate the congestion. Similarly, a smart home entertainment system can adjust the room's lighting and temperature based on the content consumed, creating a more immersive and enjoyable experience.

However, it is important to note that the personalization and customization of infotainment experiences through IoT come with certain challenges, particularly in terms of security and privacy. As these systems collect and store vast amounts of user data, there is an increased risk of data breaches and unauthorized access to sensitive information. Additionally, users may be concerned about how much their personal data is being used and shared by IoT-enabled infotainment systems. Addressing these concerns will be crucial in ensuring the continued growth and success of IoT-based infotainment solutions.

In conclusion, the Internet of Things has significantly transformed the world of infotainment, offering users highly personalized and customized experiences that cater to their unique preferences and needs. From data-driven content recommendations to context-aware adaptations, IoT-based infotainment systems are revolutionizing the way we consume and interact with content. As we continue to embrace the interconnected world of IoT, the future of infotainment promises to be more engaging, immersive, and tailored to the individual user than ever before.

Security and Privacy Concerns in IoT-based Infotainment Systems

As infotainment systems become increasingly interconnected with the Internet of Things (IoT), it is crucial to address the security and privacy concerns that arise from this integration. While IoT-based infotainment systems offer numerous benefits, such as personalized experiences and seamless integration with smart home and smart city technologies, they also present potential risks to users' personal information and overall system security. This section will explore the various security and privacy concerns associated with IoT-based infotainment systems and discuss potential solutions to mitigate these risks.

One of the primary concerns with IoT-based infotainment systems is the potential for unauthorized access to users' personal information. As these systems collect and store vast amounts of data, including users' preferences, location history, and even biometric data, it is essential to ensure this information remains secure and confidential. To address this concern, companies must implement robust data encryption and secure storage solutions, as well as adopt transparent data collection policies that inform users about the types of data being collected and how they will be used.

IoT-based infotainment systems often rely on multiple devices and networks to function seamlessly. This interconnectedness, however, also presents potential security vulnerabilities. Hackers may exploit weak points in the system to gain unauthorized access, potentially compromising the infotainment system and other connected devices and networks. To mitigate this risk, companies must prioritize device and network security by implementing strong authentication protocols, regularly updating software and firmware, and monitoring for potential threats.

As infotainment systems become more integrated with third-party applications and services, the potential for security breaches increases. Users must be cautious when granting permissions to these third-party integrations, as they may inadvertently expose their personal information or create vulnerabilities in the system. Companies should carefully vet third-party partners and ensure that they adhere to strict security and privacy standards.

A crucial aspect of addressing security and privacy concerns in IoT-based infotainment systems is educating users about potential risks and best practices for protecting their information. Companies should provide clear guidelines on how to create strong passwords, enable two-factor authentication, and regularly update software. Additionally, users should be encouraged to review and adjust privacy settings to ensure that they are comfortable with the level of data being collected and shared.

In conclusion, the integration of infotainment systems with IoT technologies presents both exciting opportunities and potential security and privacy risks. By prioritizing data protection, device and network security, and user education, companies can help ensure that users enjoy the benefits of IoT-based infotainment systems while minimizing potential threats. As the IoT continues to evolve, industry leaders need to remain vigilant in addressing these concerns and fostering a secure and privacy-conscious environment for users.

The Future of Infotainment in an Interconnected World

As we stand at the precipice of a new era in technology, the future of infotainment in an interconnected world is both exciting and full of potential. The Internet of Things (IoT) has already begun to revolutionize the way we interact with our environment, and its integration with infotainment systems promises to create a seamless and personalized experience for users across various platforms.

The evolution of infotainment systems in the IoT era has seen a rapid transformation from standalone devices to interconnected ecosystems that communicate with each other and adapt to the user's preferences. This has allowed for a more immersive and engaging experience, as users can now access a wealth of information and entertainment options at their fingertips.

The integration of infotainment with smart home and smart city technologies has further expanded the possibilities for users, allowing them to control various aspects of their environment through a single interface. This simplifies the user experience and enables more efficient use of resources, ultimately contributing to a more sustainable future.

Personalization and customization of infotainment experiences through IoT have also become a key focus for developers as they strive to create systems that cater to each user's unique preferences and needs. This has led to the development of advanced algorithms and machine learning techniques that can analyze user behavior and adapt the infotainment system accordingly, ensuring a truly tailored experience.

However, the rapid expansion of IoT-based infotainment systems has also raised concerns about security and privacy. As more devices become interconnected, the potential for cyberattacks and data breaches increases, posing a significant risk to users. It is therefore crucial for developers and industry experts to prioritize the development of robust security measures and educate users about the importance of safeguarding their personal information.

In conclusion, the future of infotainment in an interconnected world is undoubtedly bright, with the potential to transform the way we consume information and entertainment. As IoT technology continues to evolve and integrate with infotainment systems, we can expect to see even more innovative and personalized experiences for users. However, we must remain vigilant about the potential risks associated with this rapid growth and ensure that security and privacy remain at the forefront of development. By striking the right balance between innovation and security, we can look forward to a truly interconnected world where infotainment systems enrich our lives and empower us to make informed decisions.

Chapter Summary

- Infotainment, a blend of information and entertainment, has evolved significantly with the advent of the Internet of Things (IoT), offering users a more personalized and immersive experience.
- IoT has enabled the integration of infotainment systems with various connected devices and platforms, leading to the development of advanced systems such as connected cars, smart TVs, voice assistants, and wearable technology.
- The integration of infotainment with smart home and smart city technologies has enhanced user experiences by providing centralized control, voice-activated assistance, and personalized recommendations.
- IoT allows for personalization and customization of infotainment experiences by collecting and analyzing user data, enabling context-aware adaptations and data-driven content recommendations.
- Security and privacy concerns arise with IoT-based infotainment systems due to the potential for unauthorized access to users' personal information and overall system security.
- To address security and privacy concerns, companies must implement robust data encryption, secure storage solutions, strong authentication protocols, and regular software updates.

- User awareness and education are crucial in addressing security and privacy concerns, with companies providing guidelines on creating strong passwords, enabling two-factor authentication, and adjusting privacy settings.
- The future of infotainment in an interconnected world is promising, with the potential to transform the way we consume information and entertainment. However, striking the right balance between innovation and security is essential to ensure a safe and privacy-conscious environment for users.

7

EMERGING TECHNOLOGIES IN INFOTAINMENT

A futuristic infotainment system seamlessly integrated into the dashboard of a sleek, self-driving car.

Infotainment has become a vital aspect of our daily lives. It refers to the fusion of informative and entertaining content delivered through various media platforms such as television, radio, print, and digital channels. The primary goal of infotainment is to educate and inform audiences while keeping them engaged and entertained. As technology continues to evolve rapidly, the infotainment industry is undergoing a significant transformation, with emerging technologies playing a crucial role in shaping its future.

In this chapter, we will delve into the world of emerging technologies in infotainment, exploring how innovations such as augmented and virtual reality, artificial intelligence, 5G, the Internet of Things (IoT), and voice assistants are revolutionizing the way we consume and interact with content. These technologies are enhancing the overall infotainment experience and opening up new avenues for content creators and providers to engage with their audiences in more immersive and personalized ways.

The rise of augmented and virtual reality has paved the way for more interactive and immersive infotainment experiences, allowing users to step into virtual worlds or overlay digital information onto their physical surroundings. Artificial intelligence and machine learning are revolutionizing content creation, enabling the development of more personalized and engaging content that caters to individual preferences and interests. The advent of 5G and IoT is set to transform the way we access and consume infotainment, providing seamless connectivity and enabling the integration of various smart devices for a more unified experience. Voice assistants and smart speakers are also gaining prominence in the infotainment landscape, offering users a more convenient and hands-free way to access content.

As we explore these emerging technologies in infotainment, it is essential to understand their transformative potential and the opportunities they present for both content creators and consumers. By embracing these innovations, the infotainment industry can continue to evolve and thrive in an increasingly digital and connected world, offering audiences more engaging, informative, and entertaining experiences than ever before.

The Rise of Augmented and Virtual Reality in Infotainment

In recent years, the infotainment landscape has been transformed by the rapid development and adoption of augmented reality (AR) and virtual reality (VR) technologies. These cutting-edge innovations have opened up new possibilities for content creators and consumers alike, allowing for immersive and interactive experiences that were once the stuff of science fiction. In this section, we will explore the rise of AR and VR in infotainment, examining their potential to revolutionize the way we consume and engage with content.

Augmented reality refers to the overlaying of digital information onto the user's real-world environment, typically through the use of a smartphone or wearable device. This technology has found a variety of applications in the infotainment sphere, from interactive museum exhibits to location-based games like Pokémon Go. By blending the digital and physical worlds, AR has the power to enrich our experiences and provide contextually relevant information in real time.

Virtual reality, on the other hand, transports users into a fully digital environment, often through the use of a head-mounted display (HMD) or VR headset. The entertainment industry has embraced this technology, with VR experiences ranging from immersive films and documentaries to interactive video games. In addition to providing a new level of immersion, VR has the potential to democratize access to experiences that were once limited by geographical or financial constraints, such as attending a live concert or exploring a far-off destination.

One of the most significant impacts of AR and VR on infotainment is the shift from passive consumption to active engagement. These technologies encourage users to become active participants in their experiences, whether it's by interacting with digital objects in an AR app or navigating a virtual environment in VR. This increased interactivity makes content more engaging and opens up new avenues for learning and exploration.

Moreover, AR and VR have the potential to create more personalized and tailored infotainment experiences. By leveraging user data and preferences, content creators can develop experiences that cater to individual interests and needs. For example, a virtual reality travel experience could be customized to highlight specific landmarks or activities based on the user's preferences, while an augmented reality museum tour could adapt its content to the user's level of expertise or interest in a particular subject.

In conclusion, the rise of augmented and virtual reality in infotainment marks a significant shift in the way we consume and engage with content. These emerging technologies have the power to transform passive experiences into interactive and immersive adventures while also offering the potential for more personalized and tailored content. As AR and VR continue to evolve and become more accessible, we can expect to see even more innovative and engaging infotainment experiences in the years to come.

Artificial Intelligence and Machine Learning: Revolutionizing Content Creation

In recent years, the infotainment industry has experienced a significant shift in the way content is created and consumed, thanks to the rapid advancements in artificial intelligence (AI) and machine learning (ML) technologies. These cutting-edge technologies have not only transformed the way we access and interact with information but also revolutionized the content creation process itself. In this section, we will delve into the impact of AI and ML on content creation and explore how these technologies are shaping the future of infotainment.

One of the most significant ways AI and ML have impacted content creation is by enabling the generation of personalized content. By analyzing user preferences, behaviors, and interests, AI-powered algorithms can create tailor-made content that caters to the specific needs and tastes of individual users. This level of personalization has led to a more engaging and immersive infotainment experience, as users are now presented with highly relevant and appealing content.

AI and ML have also paved the way for automated content production, which has the potential to revolutionize the way content is created and distributed. With the help of natural language processing (NLP) and natural language generation (NLG) technologies, AI-powered systems can now generate news articles, social media posts, and even creative writing pieces with minimal human intervention. This saves time and resources and allows content creators to focus on more strategic and creative aspects of their work.

The integration of AI and ML in content creation has also led to significant improvements in visual and audio content. For instance, AI-powered image and video editing tools can now automatically enhance the quality of visuals, remove unwanted elements, and even generate realistic images from scratch. Similarly, AI-driven audio processing tools can analyze and modify sound to create immersive audio experiences, such as 3D audio and adaptive soundtracks.

Another notable application of AI and ML in content creation is predictive content analytics. By analyzing historical data and user behavior patterns, AI-powered systems can predict which content is likely to resonate with users and generate higher engagement. This enables content creators to make data-driven decisions and optimize their content strategies for maximum impact.

While AI and ML have made significant strides in content creation, they are not meant to replace human creativity. Instead, these technologies can be seen as powerful tools that complement and enhance human creativity. By leveraging AI and ML, content creators can automate repetitive tasks, gain valuable insights, and focus on the more creative aspects of their work. This collaborative approach between AI and human creators has the potential to unlock new possibilities and push the boundaries of infotainment content.

In conclusion, the integration of artificial intelligence and machine learning in content creation has brought about a paradigm shift in the infotainment industry. By enabling personalized content generation, automating content production, enhancing visual and audio content, and providing predictive content analytics, AI and ML have revolutionized the way content is created and consumed. As these technologies continue to evolve, we can expect even more innovative and immersive infotainment experiences in the years to come.

The Impact of 5G and IoT on Infotainment Experiences

The world of infotainment is continuously evolving, and the introduction of 5G and the Internet of Things (IoT) is set to revolutionize the way we consume and interact with content. This section will explore how these emerging technologies transform infotainment experiences, enabling faster connectivity, seamless integration, and personalized content delivery.

The Power of 5G Connectivity

5G, the fifth generation of mobile networks, promises to deliver lightning-fast internet speeds, ultra-low latency, and increased network capacity. This technological leap will profoundly impact infotainment experiences, as it enables streaming of high-quality content without buffering or lag. Imagine watching a live sports event or concert in 4K resolution on your mobile device without any interruptions or delays – that's the power of 5G connectivity.

Moreover, 5G will enable the seamless integration of augmented and virtual reality experiences into infotainment platforms. With its low latency and high-speed data transfer capabilities, 5G will allow users to enjoy immersive AR and VR content without any performance issues, further blurring the lines between the digital and physical worlds.

The Internet of Things (IoT) and Infotainment

The Internet of Things (IoT) refers to the interconnection of everyday objects, such as appliances, vehicles, and wearable devices, through the internet. This technology has the potential to revolutionize the infotainment landscape by providing seamless integration and personalization of content across multiple devices.

For instance, IoT-enabled smart TVs can automatically recommend content based on your viewing habits and preferences, while smart speakers can curate playlists based on your mood or the time of day. In the automotive industry, connected cars can provide real-time traffic updates, navigation assistance, and personalized entertainment options, making your daily commute more enjoyable and efficient.

The Synergy of 5G and IoT in Infotainment

The combination of 5G and IoT technologies will create a powerful synergy that will redefine the infotainment experience. With 5G's high-speed connectivity and IoT's seamless integration capabilities, users can expect a more immersive, personalized, and interactive infotainment experience.

For example, imagine attending a live concert through a virtual reality headset, where you can choose your vantage point, interact with other attendees, and even influence the performance in real time. Or picture a smart home where your IoT devices work together to create a personalized infotainment experience, adjusting lighting, temperature, and audio settings to match the content you're consuming.

The impact of 5G and IoT on infotainment experiences cannot be overstated. These emerging technologies will enhance how we consume content and create new opportunities for content creators and service providers to deliver innovative and immersive experiences. As 5G and IoT continue to gain traction, we can expect a future where infotainment is more engaging, personalized, and integrated into our daily lives than ever before.

The Future of Voice Assistants and Smart Speakers in Infotainment

Rapid technological advancements have paved the way for voice assistants and smart speakers to become an integral part of our daily lives. These devices, powered by artificial intelligence, are revolutionizing the way we interact with technology and consume information. As the infotainment industry continues to evolve, voice assistants and smart speakers are poised to play a significant role in shaping the future of this domain.

Voice assistants, such as Amazon's Alexa, Apple's Siri, and Google Assistant, are becoming increasingly sophisticated, capable of understanding natural language and providing personalized responses. This has led to a surge in the popularity of smart speakers, which serve as a convenient and hands-free medium for accessing infotainment content. As these technologies continue to advance, we can expect them to become even more integrated into our infotainment experiences.

One of the most significant impacts of voice assistants and smart speakers on the infotainment industry is the shift toward audio-based content. Podcasts, audiobooks, and news briefings are gaining popularity as users seek convenient and engaging ways to consume information. This trend is expected to continue, with voice assistants and smart speakers becoming the primary medium for accessing and discovering audio content.

Moreover, voice assistants and smart speakers are transforming the way we interact with our devices, making them more natural and intuitive. As a result, content creators and providers need to adapt their strategies to cater to this new mode of interaction. This includes optimizing their content for voice search, creating voice-activated applications, and developing interactive experiences that leverage the unique capabilities of voice assistants.

Another exciting development in voice assistants and smart speakers is the potential for personalized and context-aware infotainment experiences. As these devices become more intelligent, they will be able to understand user preferences, habits, and contexts, allowing them to deliver tailored content that is both relevant and engaging. This level of personalization will enhance the user experience and create new opportunities for targeted advertising and content monetization.

In addition to personalization, the future of voice assistants and smart speakers in infotainment will also be shaped by the integration of other emerging technologies. For instance, the combination of voice assistants with augmented and virtual reality could lead to immersive and interactive experiences that redefine the way we consume infotainment content. Similarly, the convergence of voice assistants with IoT devices and 5G networks could enable seamless and interconnected infotainment experiences across various platforms and devices.

In conclusion, the future of voice assistants and smart speakers in infotainment is incredibly promising, with the potential to transform the way we access, discover, and engage with content. As these technologies continue to advance, we expect them to become even more integral to our infotainment experiences, driving innovation and creating new opportunities for content creators and consumers. Embracing these changes and adapting to the evolving landscape will be crucial for the infotainment industry to thrive in the era of voice assistants and smart speakers.

The Transformative Potential of Emerging Technologies in Infotainment

As we have explored throughout this chapter, the world of infotainment is undergoing a rapid and exciting transformation, thanks to the integration of emerging technologies. From augmented and virtual reality to artificial intelligence and machine learning, these advancements are revolutionizing the way we create, consume, and interact with content. As we look to the future, it is clear that these technologies hold immense potential to reshape the infotainment landscape, offering new opportunities for innovation and growth.

One of the most significant impacts of these emerging technologies is the democratization of content creation. With tools like AI and machine learning, creators can now develop personalized, immersive experiences that cater to individual preferences and needs. This enhances the overall quality of infotainment and fosters a more inclusive and diverse ecosystem where a wider range of voices and perspectives can be heard.

Furthermore, the advent of 5G and IoT technologies can revolutionize how we access and engage with infotainment. As these technologies become more widespread, we can expect to see a seamless integration of infotainment experiences into our daily lives, with content being delivered faster and more efficiently than ever before. This will enable us to stay connected and informed, no matter where we are or what we are doing.

Voice assistants and smart speakers are another area of growth in the infotainment sector, as they offer a convenient and accessible way for users to access content. As these devices become more sophisticated and capable of understanding natural language, we can expect to see a shift towards voice-based interactions, making infotainment experiences more intuitive and user-friendly.

In conclusion, the transformative potential of emerging technologies in infotainment is undeniable. As we continue to push the boundaries of what is possible, we can look forward to a future where infotainment is more immersive, personalized, and accessible than ever before. By embracing these innovations, we can unlock new opportunities for growth and development, ensuring that infotainment remains a vital and dynamic part of our lives for years to come.

Chapter Summary

- Infotainment, the fusion of informative and entertaining content, is undergoing a significant transformation due to emerging technologies such as augmented and virtual reality, artificial intelligence, 5G, IoT, and voice assistants.
- Augmented and virtual reality are revolutionizing infotainment by providing more interactive and immersive experiences, allowing users to actively engage with content and enabling personalized experiences based on individual preferences.
- Artificial intelligence and machine learning are transforming content creation by enabling personalized content generation, automating content production,

enhancing visual and audio content, providing predictive content analytics, and fostering collaboration between AI and human creators.

- The advent of 5G and IoT technologies is set to transform infotainment experiences by providing faster connectivity, seamless integration, and personalized content delivery across multiple devices.
- Voice assistants and smart speakers are becoming increasingly integral to infotainment experiences, driving a shift towards audio-based content and transforming the way users interact with devices.
- Personalization and context-aware infotainment experiences will be a key focus in the future, with voice assistants and smart speakers becoming more intelligent and capable of understanding user preferences, habits, and contexts.
- The integration of emerging technologies such as augmented and virtual reality, IoT devices, and 5G networks with voice assistants and smart speakers will enable seamless and interconnected infotainment experiences across various platforms and devices.
- The transformative potential of emerging technologies in infotainment is immense, offering new opportunities for innovation, growth, and the democratization of content creation, ultimately leading to more immersive, personalized, and accessible infotainment experiences.

8

CYBERSECURITY AND PRIVACY IN INFOTAINMENT SYSTEMS

A futuristic car infotainment system with advanced cybersecurity features.

In today's hyper-connected world, infotainment systems have become an integral part of our daily lives. These systems, which combine information and entertainment, are found in various forms, such as in-car entertainment systems, smart TVs, and streaming platforms. As the popularity of infotainment systems continues to grow, so does the need for robust cybersecurity and privacy measures to protect users and their data.

Cybersecurity and privacy in infotainment systems are crucial aspects that ensure the safety and security of users' personal information, as well as the proper functioning of the systems themselves. With the increasing amount of sensitive data being stored and transmitted through these systems, the risk of cyberattacks and data breaches has never been higher. This chapter aims to provide an overview of the importance of cybersecurity and privacy in infotainment systems, the common threats and vulnerabilities faced by these systems, and the best practices for ensuring their security and privacy.

As infotainment systems become more sophisticated and interconnected, they also become more vulnerable to cyber threats. Hackers and cybercriminals are constantly looking for ways to exploit vulnerabilities in these systems, which can lead to unauthorized access, data theft, and even physical harm to users. For instance, a hacker could potentially gain control of a vehicle's infotainment system and manipulate its functions, putting the driver and passengers at risk.

Moreover, privacy concerns have become increasingly relevant in the context of infotainment systems. Users often provide personal information, such as their location, preferences, and payment details, to access various features and services offered by these systems. If not properly protected, this data can be misused by malicious actors for identity theft, financial fraud, or other nefarious purposes.

In light of these risks, manufacturers, developers, and users of infotainment systems need to prioritize cybersecurity and privacy. By implementing strong security measures and adhering to industry standards and regulations, we can ensure that infotainment systems remain a safe and enjoyable part of our lives. In the following sections, we will delve deeper into the various aspects of cybersecurity and privacy in infotainment systems, providing insights and recommendations for a secure and private user experience.

The Growing Importance of Cybersecurity in Infotainment Systems

As infotainment systems continue to evolve and become more integrated into our daily lives, the importance of cybersecurity in these systems cannot be overstated. With the rapid advancement of technology, infotainment systems have transformed from simple radio and CD players to sophisticated, connected devices that offer a wide range of services such as navigation, communication, and entertainment. These systems are now essential to modern vehicles, smart homes, and personal devices, making them an attractive target for cybercriminals.

There are several factors contributing to the growing importance of cybersecurity in infotainment systems:

- **Increased Connectivity:** The integration of infotainment systems with the internet and other connected devices has opened up new avenues for potential cyberattacks. Hackers can exploit vulnerabilities in these systems to gain unauthorized access to sensitive information, manipulate system functionality, or even take control of the device.
- **Data Collection and Storage:** Modern infotainment systems collect and store a vast
- amount of personal data, such as user preferences, location history, and communication logs. This data can be highly valuable to cybercriminals, who may use it for identity theft, targeted advertising, or other malicious purposes.
- **Complexity of Systems:** The complexity of infotainment systems has increased exponentially in recent years with the addition of new features, applications, and interfaces. This complexity makes it more challenging to identify and

address potential security vulnerabilities, leaving the systems more susceptible to cyberattacks.

- **User Trust and Expectations:** As users become more reliant on infotainment systems for their daily needs, they expect these systems to be secure and protect their privacy. Any security or privacy breach can lead to a loss of trust in the system and its manufacturer, resulting in reputational damage and potential legal consequences.
- **Interconnectivity with Critical Systems:** In the case of automotive infotainment systems, the increasing integration with other vehicle systems, such as engine control and safety features, raises the stakes for cybersecurity. A successful cyberattack on an infotainment system could potentially compromise the safety and functionality of the entire vehicle.

In light of these factors, it is crucial for manufacturers, developers, and users of infotainment systems to prioritize cybersecurity and privacy. By understanding the risks and implementing robust security measures, we can ensure that infotainment systems continue to provide the convenience and entertainment we have come to expect while safeguarding our personal information and safety.

Common Cybersecurity Threats and Vulnerabilities in Infotainment Systems

As infotainment systems become more advanced and interconnected, they also become more susceptible to various cybersecurity threats and vulnerabilities. In this section, we will explore some of the most common threats and vulnerabilities that infotainment systems face today, and how they can potentially impact users' privacy and security.

One of the primary concerns in infotainment systems is unauthorized access to sensitive data. This can occur through various means, such as hacking into the system, exploiting software vulnerabilities, or intercepting data transmissions. Data breaches can lead to the exposure of personal information, such as user preferences, location data, and even payment details, which can then be used for identity theft, fraud, or other malicious activities.

Malware and ransomware attacks are becoming increasingly common in infotainment systems. Cybercriminals can use these attacks to gain control of a user's infotainment system, lock them out, or even demand a ransom to restore access. In some cases, malware can also be used to track a user's location, monitor their activities, or steal sensitive information.

As infotainment systems become more complex, they also become more prone to software vulnerabilities and exploits. Hackers can exploit these vulnerabilities to gain unauthorized access to the system, manipulate its functions, or even take control of the vehicle itself. In some cases, these vulnerabilities can be discovered and exploited by cybercriminals before manufacturers have the chance to issue a patch or update, leaving users at risk.

Many infotainment systems rely on wireless communication technologies, such as Wi-Fi, Bluetooth, and cellular networks, to provide connectivity and functionality. However, these wireless connections can also be vulnerable to eavesdropping, interception, or tampering. This can lead to the exposure of sensitive data, unauthorized access to the infotainment system, or even the spread of malware between connected devices.

As infotainment systems become more open and customizable, users often have the option to install third-party applications and services. While this can provide additional functionality and convenience, it can also introduce new security risks. Some third-party applications may contain vulnerabilities or malicious code, which can compromise the security of the infotainment system and potentially lead to data breaches or other cyberattacks.

In conclusion, the growing complexity and connectivity of infotainment systems have led to an increased risk of cybersecurity threats and vulnerabilities. Therefore, manufacturers, developers, and users need to be aware of these risks and take appropriate measures to protect their systems and personal information. The following section will discuss best practices for ensuring privacy and security in infotainment systems.

Best Practices for Ensuring Privacy and Security in Infotainment Systems

As infotainment systems continue to evolve and become more integrated into our daily lives, it is crucial to prioritize privacy and security. This section will outline some best practices that can help ensure the safety and privacy of users while enjoying the benefits of infotainment systems.

Regular Software Updates: One of the most effective ways to protect infotainment systems from cyber threats is to keep the software up-to-date. Manufacturers should release regular updates to fix any security vulnerabilities and improve the overall performance of the system. Users should also be proactive in installing these updates as soon as they become available.

Strong Authentication Measures: Implementing strong authentication measures, such as multi-factor authentication or biometric identification, can significantly reduce the risk of unauthorized access to infotainment systems. This is particularly important for systems that store sensitive personal information or provide access to financial transactions.

Data Encryption: Encrypting data stored on infotainment systems can help protect user privacy and prevent unauthorized access to sensitive information. Manufacturers should use robust encryption algorithms and ensure that encryption keys are securely managed.

Secure Communication Channels: Infotainment systems often communicate with external devices and networks, making it essential to secure these communication channels. This can be achieved by using secure protocols, such as HTTPS and SSL/TLS, and regularly updating security certificates.

Privacy by Design: Manufacturers should adopt a privacy-by-design approach when developing infotainment systems. This means considering privacy and security implications from the initial design stages and incorporating appropriate safeguards throughout the development process.

User Education and Awareness: Users play a critical role in maintaining the privacy and security of infotainment systems. Manufacturers and service providers should provide clear guidance on best practices for using these systems safely and securely. This may include instructions on setting strong passwords, enabling security features, and recognizing potential phishing attempts.

Third-Party Security Audits: Regular third-party security audits can help identify potential vulnerabilities in infotainment systems and ensure that manufacturers adhere to industry best practices. These audits should be conducted by reputable cybersecurity firms with experience in the infotainment industry.

Incident Response Planning: Manufacturers should have a well-defined incident response plan in place to address any potential security breaches or privacy violations. This plan should outline the steps to be taken in the event of an incident, including notifying affected users, mitigating the impact, and preventing future occurrences.

By following these best practices, manufacturers and users can work together to create a safer and more secure infotainment ecosystem. As technology continues to advance, it is essential to remain vigilant and proactive in addressing potential cybersecurity and privacy risks. By doing so, we can ensure that infotainment systems continue to provide valuable and enjoyable experiences while protecting the privacy and security of users.

The Role of Legislation and Industry Standards in Protecting Infotainment Systems

As infotainment systems continue to evolve and become more sophisticated, the need for robust cybersecurity measures and privacy protection becomes increasingly critical. In this section, we will explore the role of legislation and industry standards in safeguarding infotainment systems and ensuring users can enjoy these technologies without compromising their security and privacy.

Legislation is crucial in setting the legal framework for cybersecurity and privacy protection in infotainment systems. Governments around the world have recognized the importance of securing these systems and have enacted laws and regulations to ensure that manufacturers and service providers adhere to certain security and privacy standards. For example, the European Union's General Data Protection Regulation (GDPR) has significantly impacted how companies handle user data, including data collected and processed by infotainment systems. In the United States, the Federal Trade Commission (FTC) enforces privacy and data security requirements for connected devices, including infotainment systems.

Industry standards also play a vital role in shaping the security and privacy landscape for infotainment systems. Organizations like the International Organization for Standardization (ISO) and the Society of Automotive Engineers (SAE) have developed guidelines and best practices for securing automotive systems, including infotainment systems. For instance, the ISO/SAE 21434 standard provides a comprehensive framework for addressing cybersecurity risks in the automotive industry, covering aspects such as risk management, vulnerability assessment, and incident response.

These industry standards serve as a benchmark for manufacturers and service providers, ensuring they follow best practices in securing their infotainment systems. By adhering to these standards, companies can demonstrate their commitment to cybersecurity and privacy, thereby building trust with consumers and regulatory authorities.

In addition to legislation and industry standards, collaboration between various stakeholders is essential in protecting infotainment systems. This includes cooperation between manufacturers, suppliers, software developers, and cybersecurity experts to share information about potential threats and vulnerabilities, as well as to develop and implement effective security measures. Public-private partnerships, such as the Automotive Information Sharing and Analysis Center (Auto-ISAC), facilitate this collaboration by providing a platform for sharing cybersecurity information and best practices among industry stakeholders.

In conclusion, legislation and industry standards play a pivotal role in ensuring the security and privacy of infotainment systems. By establishing a legal framework and setting benchmarks for best practices, these measures help to protect users from potential cybersecurity threats and privacy breaches. As infotainment systems continue to advance and become more interconnected, it is essential for all stakeholders to work together to stay ahead of emerging risks and safeguard the future of these technologies.

The Future of Cybersecurity and Privacy in Infotainment Systems

As we reach the end of this chapter, it is important to reflect on the significance of cybersecurity and privacy in the rapidly evolving world of infotainment systems. The integration of technology into our daily lives has brought about a plethora of opportunities for entertainment, communication, and convenience. However, with these advancements come the inevitable challenges of ensuring the security and privacy of users' data and personal information.

The future of cybersecurity and privacy in infotainment systems will be shaped by the continuous development of new technologies and the growing awareness of the potential risks associated with their use. As the industry moves forward, it is crucial for manufacturers, developers, and users alike to prioritize the protection of sensitive information and the prevention of cyber threats.

One of the key factors in achieving this goal is the implementation of robust security measures and best practices throughout the design and development process. By incorporating security features such as encryption, authentication, and regular software updates, manufacturers can significantly reduce the likelihood of cyberattacks and data breaches.

Moreover, the role of legislation and industry standards will continue to be critical in shaping the future of cybersecurity and privacy in infotainment systems. Governments and regulatory bodies must work together to establish comprehensive guidelines and requirements that promote the development of secure and privacy-respecting technologies. This collaborative effort will not only help to protect users' data but also foster trust and confidence in the infotainment industry as a whole.

In addition to these measures, the future of cybersecurity and privacy in infotainment systems will also be influenced by users' growing awareness and involvement. As consumers become more educated about the potential risks associated with their devices, they will be better equipped to make informed decisions about the products they use and the security measures they implement. This increased vigilance will ultimately contribute to the overall security and privacy of infotainment systems.

In conclusion, the future of cybersecurity and privacy in infotainment systems is a dynamic and multifaceted landscape that will be shaped by the ongoing collaboration between manufacturers, developers, governments, and users. By prioritizing security and privacy throughout the design, development, and implementation process, the industry can continue to innovate and evolve while ensuring the protection of users' data and personal information. As we move forward, it is essential to remain vigilant and proactive in addressing the ever-changing challenges and opportunities presented by the world of infotainment technology.

Chapter Summary

- The growing importance of cybersecurity in infotainment systems is due to increased connectivity, data collection and storage, complexity of systems, user trust and expectations, and interconnectivity with critical systems.
- Common cybersecurity threats and vulnerabilities in infotainment systems include unauthorized access and data breaches, malware and ransomware attacks, software vulnerabilities and exploits, insecure wireless communications, and third-party applications and services.

- Best practices for ensuring privacy and security in infotainment systems include regular software updates, strong authentication measures, data encryption, secure communication channels, privacy by design, user education and awareness, third-party security audits, and incident response planning.
- Legislation and industry standards play a crucial role in setting the legal framework for cybersecurity and privacy protection in infotainment systems, with examples such as the European Union's General Data Protection Regulation (GDPR) and the ISO/SAE 21434 standard.
- Collaboration between various stakeholders, including manufacturers, suppliers, software developers, and cybersecurity experts, is essential in protecting infotainment systems and sharing information about potential threats and vulnerabilities.
- Public-private partnerships, such as the Automotive Information Sharing and Analysis Center (Auto-ISAC), facilitate collaboration by providing a platform for sharing cybersecurity information and best practices among industry stakeholders.
- The future of cybersecurity and privacy in infotainment systems will be shaped by the continuous development of new technologies, growing awareness of potential risks, and the implementation of robust security measures and best practices throughout the design and development process.
- The ongoing collaboration between manufacturers, developers, governments, and users is essential for prioritizing security and privacy in infotainment systems, fostering trust and confidence in the industry, and ensuring the protection of users' data and personal information.

9

INFOTAINMENT AND ITS IMPACT ON SOCIETY

A futuristic smart city, where various Internet of Things (IoT) devices seamlessly communicate with each other.

In today's fast-paced and interconnected world, the media landscape has undergone a significant transformation. The traditional boundaries between news and entertainment have become increasingly blurred, giving rise to a new phenomenon known as infotainment. This term refers to the fusion of informative content with elements of entertainment, often presented in a sensationalized or attention-grabbing manner. As the demand for engaging and easily digestible content continues to grow, infotainment has become a prominent fixture in modern media, shaping the way we consume information and influencing our perceptions of the world around us.

The rise of infotainment can be attributed to several factors, including the proliferation of media outlets, the advent of the 24-hour news cycle, and the increasing influence of social media and technology. With an ever-growing number of channels and platforms vying for audience attention, media organizations have been compelled to adopt new strategies to attract and retain viewers. Infotainment has become a popular solution, offering a more engaging and accessible alternative to traditional news formats.

This shift towards infotainment has been further fueled by the rapid advancement of technology and the widespread adoption of social media. As smartphones and other devices have become increasingly ubiquitous, consumers now have instant access to a vast array of information and entertainment options at their fingertips. This has led to a significant change in audience expectations, with many people now seeking out content that is not only informative but also entertaining and shareable. In response, media organizations have increasingly turned to infotainment to cater to these evolving preferences.

While the rise of infotainment has undoubtedly transformed the way we consume information, it has also raised important questions about the impact of this phenomenon on society. As the lines between news and entertainment continue to blur, concerns have been raised about the potential consequences for public opinion, political discourse, and the ethical responsibilities of media providers. In this chapter, we will explore these issues in greater depth, examining the various ways in which infotainment has shaped our world and considering the implications for the future of media and society.

The Blurring Lines Between News and Entertainment

In recent years, the distinction between news and entertainment has become increasingly blurred, giving rise to the phenomenon of infotainment. This fusion of information and entertainment has led to a shift in the way people consume and perceive news, as well as the manner in which media outlets present it. In this section, we will delve into the factors contributing to this blurring of lines and the implications it has on society.

One of the primary reasons for the blurring of lines between news and entertainment is the fierce competition among media outlets for viewership and advertising revenue. In an attempt to attract a larger audience, news organizations have increasingly adopted sensationalist tactics, focusing on stories that are more entertaining and attention-grabbing rather than those of significant public interest. This has led to a shift in news content, with a greater emphasis on celebrity gossip, scandals, and human-interest stories, often at the expense of in-depth reporting on important issues.

Another contributing factor is the rise of the 24-hour news cycle, which has created a constant demand for fresh content. In order to fill this void, news organizations often resort to presenting stories in a more entertaining and engaging manner, with flashy visuals, dramatic storytelling, and even humorous commentary. This approach not only blurs the lines between news and entertainment but also has the potential to trivialize important issues and undermine the credibility of news organizations.

The influence of infotainment is also evident in the increasing prominence of opinion-driven news programming, where the focus is on providing commentary and analysis rather than objective reporting. This format often features charismatic personalities who express strong opinions and engage in heated debates, making for compelling television. However, this approach can contribute to the polarization of public opinion, as viewers are more likely to be exposed to biased perspectives that align with their own beliefs.

Moreover, the advent of social media and technology has played a significant role in the blurring of lines between news and entertainment. Platforms like Facebook and Twitter have become primary sources of news for many people, who often encounter news stories mixed in with entertaining content, such as memes, videos, and personal updates from friends. This blending of news and entertainment on social media platforms can make it difficult for users to distinguish between factual reporting and opinion, leading to the spread of misinformation and the erosion of trust in traditional news sources.

In conclusion, the blurring of lines between news and entertainment has significant implications for society, as it affects the way people consume and perceive news, as well as the manner in which media outlets present it. As infotainment continues to gain prominence in modern media, it is crucial for both consumers and providers of news to be aware of the potential consequences and to strive for a balance between engaging content and responsible journalism.

The Influence of Infotainment on Public Opinion and Political Discourse

The rise of infotainment has significantly impacted public opinion and political discourse. As the lines between news and entertainment continue to blur, the way people consume information and form opinions has evolved. This section will explore the ways in which infotainment influences public opinion and political discourse, and the potential consequences of this shift.

One of the most apparent effects of infotainment is the simplification of complex issues. Infotainment often presents information in an easily digestible and entertaining manner, which can lead to oversimplification. This can result in the public forming opinions based on incomplete or inaccurate information. For example, a news segment that focuses on the personal life of a political figure rather than their policies may lead viewers to form opinions based on superficial factors rather than the substance of their political beliefs.

Infotainment has also contributed to the rise of sensationalism in news reporting. In an effort to attract viewers and generate revenue, infotainment providers often prioritize stories that are shocking or emotionally charged. This can lead to an overemphasis on scandal and controversy, which may distort public opinion and overshadow more important issues. The focus on sensationalism can also contribute to a more polarized political climate, as people are exposed to extreme viewpoints and inflammatory rhetoric.

The influence of infotainment on political discourse is further exacerbated by the role of social media and technology. As people increasingly rely on social media platforms for news and information, they are exposed to a constant stream of infotainment content. This can create echo chambers, where individuals are only exposed to information that aligns with their beliefs and opinions. This phenomenon can lead to increased political polarization and a breakdown in civil discourse as people become less willing to engage with opposing viewpoints.

Moreover, the prevalence of infotainment on social media platforms can contribute to the spread of misinformation and fake news. As infotainment providers prioritize engagement and virality, they may be more likely to share misleading or factually incorrect content. This can have serious consequences for public opinion and political discourse, as people base their opinions on false information.

In conclusion, the influence of infotainment on public opinion and political discourse is multifaceted and complex. The simplification of complex issues, the rise of sensationalism, and the role of social media and technology have all contributed to a shift in the way people consume information and form opinions. As infotainment continues to grow in popularity, it is crucial for both consumers and providers to be aware of the potential consequences and to engage in responsible and ethical media consumption.

The Role of Social Media and Technology in the Spread of Infotainment

The rapid advancement of technology and the widespread use of social media platforms have played a significant role in the proliferation of infotainment. In this digital age, people are constantly connected to the internet, seeking instant gratification and easy access to information. As a result, traditional media outlets have had to adapt to the changing landscape, often incorporating infotainment elements into their content to maintain audience engagement.

Social media platforms, such as Facebook, Twitter, and Instagram, have become powerful tools for the dissemination of infotainment. These platforms allow users to share and consume content at an unprecedented rate, often blurring the lines between news and entertainment. The algorithms used by these platforms prioritize content that generates high levels of engagement, which often means that sensationalized, emotionally charged, or controversial infotainment stories are more likely to appear in users' feeds.

This phenomenon has led to the rise of "clickbait" headlines and sensationalized content, designed to capture users' attention and encourage them to share the content with their networks. In turn, this has created an environment where infotainment thrives, as it is often more engaging and shareable than traditional news content.

Moreover, the ease with which users can create and share content on social media has given rise to a new generation of infotainment providers. Influencers, bloggers, and even ordinary users can now create content that reaches millions of people, often without the need for traditional journalistic credentials or editorial oversight. This democratization of content creation has led to an influx of infotainment, as users are drawn to the entertaining and easily digestible nature of such content.

In addition to social media, technology has also facilitated the spread of infotainment by developing new content formats and platforms. Podcasts, streaming services, and mobile applications have all contributed to the growth of infotainment by providing users with on-demand access to content that combines news and entertainment. These platforms have also given rise to a new breed of infotainment providers who are able to reach large audiences without the constraints of traditional media outlets.

While the role of social media and technology in the spread of infotainment has undoubtedly increased its reach and influence, it has also raised concerns about the quality and accuracy of the information being consumed. The ease with which false or misleading information can be shared and spread on social media platforms has led to the rise of "fake news" and misinformation, which can have serious consequences for public opinion and political discourse.

In conclusion, the role of social media and technology in the spread of infotainment has been both a driving force behind its growth and a source of concern for its potential impact on society. As infotainment continues to evolve and adapt to the changing media landscape, both content creators and consumers need to be aware of the ethical implications and responsibilities associated with this form of media.

The Ethical Implications and Responsibilities of Infotainment Providers

As infotainment continues to gain prominence in modern media, it is crucial to examine the ethical implications and responsibilities that come with providing this type of content. Infotainment providers, including television networks, online platforms, and social media influencers, must navigate a complex landscape of ethical considerations to ensure that they are not compromising the integrity of information or contributing to the spread of misinformation.

One of the primary ethical concerns surrounding infotainment is the potential for sensationalism and the prioritization of entertainment value over factual accuracy. In the pursuit of higher ratings and increased engagement, infotainment providers may be tempted to exaggerate or distort the truth to create more compelling content. This can lead to the dissemination of misleading or false information, which can have serious consequences for public understanding and decision-making. To uphold ethical standards, infotainment providers must prioritize accuracy and transparency, ensuring that their content is grounded in verifiable facts and that any potential biases or inaccuracies are clearly acknowledged.

Another ethical consideration for infotainment providers is the potential for the trivialization of important issues. By presenting complex topics in a simplified and entertaining manner, infotainment can sometimes reduce the seriousness of these issues and contribute to a lack of public engagement with critical social, political, and environmental challenges. To address this concern, infotainment providers should strive to balance entertainment value with educational content, ensuring that their audience is informed and engaged.

The role of advertising and commercial interests in infotainment also raises ethical questions. Infotainment providers must be cautious not to allow the pursuit of profit to compromise the quality and integrity of their content. This includes avoiding sensationalism, click-bait tactics, and the promotion of products or services that may be harmful or misleading. To maintain ethical standards, infotainment providers should be transparent about their financial relationships and ensure that any sponsored content is clearly labeled as such.

In addition to these ethical considerations, infotainment providers have a responsibility to promote media literacy among their audience. As the lines between news and entertainment continue to blur, consumers must be equipped with the skills to critically evaluate the information they encounter. Infotainment providers can contribute to this effort by encouraging critical thinking, fostering open dialogue, and providing resources for further learning.

In conclusion, the ethical implications and responsibilities of infotainment providers are multifaceted and complex. By prioritizing accuracy, transparency, and media literacy, these providers can help to ensure that infotainment serves as a valuable and responsible source of information and entertainment for society. As the landscape of media continues to evolve, it is essential that infotainment providers remain vigilant in upholding these ethical standards and fostering a media environment that is both engaging and informative.

The Future of Infotainment and its Effects on Society

As we have explored throughout this chapter, infotainment has become an increasingly prevalent and influential force in modern media. The blending of news and entertainment has led to a shift in the way people consume information, with many prioritizing entertainment value over factual accuracy. This phenomenon has significantly impacted public opinion, political discourse, and the ethical responsibilities of media providers. In this conclusion, we will consider the future of infotainment and its potential effects on society.

The rise of infotainment can be attributed to various factors, including the rapid advancement of technology and the widespread use of social media platforms. As these trends continue to evolve, infotainment will likely become even more deeply ingrained in our daily lives. The proliferation of smartphones, tablets, and other digital devices has made it easier than ever for people to access and share content, and this trend shows no signs of slowing down. As a result, we can expect infotainment to become even more pervasive and influential in the coming years.

One potential outcome of this increased reliance on infotainment is a further blurring of the lines between news and entertainment. As audiences continue to prioritize engaging content over objective reporting, media providers may feel compelled to cater to these preferences by producing more sensationalized and opinion-driven content. This could lead to a decline in the quality and accuracy of news reporting, with potentially serious consequences for public opinion and political discourse.

On the other hand, the growing influence of infotainment could also lead to positive developments in the media landscape. For example, the increased competition among content providers may encourage innovation and creativity in the way news is presented and disseminated. This could result in new formats and platforms that offer audiences a more engaging and informative experience, ultimately contributing to a more informed and engaged public.

In order to mitigate the potential negative effects of infotainment on society, it is crucial for media providers to recognize and uphold their ethical responsibilities. This includes ensuring that news content is accurate, balanced, and transparent, even when it is presented in an entertaining format. Additionally, media providers should strive to educate their audiences about the importance of critical thinking and media literacy, empowering them to discern between fact and fiction and make informed decisions based on reliable information.

In conclusion, the future of infotainment and its effects on society will largely depend on the choices made by both media providers and consumers. By prioritizing ethical reporting and promoting media literacy, we can harness the power of infotainment to create a more informed and engaged public while minimizing its potential negative consequences. As the lines between news and entertainment continue to blur, all stakeholders need to remain vigilant and proactive in navigating this complex and ever-evolving media landscape.

Chapter Summary

- Infotainment, a fusion of information and entertainment, has become a prominent fixture in modern media due to factors such as the proliferation of media outlets, the 24-hour news cycle, and the influence of social media and technology.
- The blurring lines between news and entertainment have led to a shift in news content, with a greater emphasis on sensationalism, celebrity gossip, and human-interest stories, often at the expense of in-depth reporting on important issues.
- Infotainment significantly impacts public opinion and political discourse, as it can lead to the simplification of complex issues, the rise of sensationalism, and the creation of echo chambers on social media platforms.
- Social media and technology have played a significant role in the spread of infotainment, with platforms like Facebook

and Twitter becoming primary news sources for many people, often blurring the lines between factual reporting and opinion.

- The ethical implications and responsibilities of infotainment providers include prioritizing accuracy and transparency, avoiding sensationalism and clickbait tactics, and promoting media literacy among their audience.
- The future of infotainment is likely to see a further blurring of the lines between news and entertainment, with potential consequences for public opinion and political discourse.
- To mitigate the potential negative effects of infotainment on society, media providers must recognize and uphold their ethical responsibilities, ensuring that news content is accurate, balanced, and transparent.
- The future of infotainment and its effects on society will largely depend on the choices made by both media providers and consumers, with a focus on ethical reporting and promoting media literacy to create a more informed and engaged public.

THE INFOTAINMENT REVOLUTION

As we reach the end of our journey exploring the fascinating world of infotainment, we must reflect on the insights and knowledge we have gained throughout this book. The infotainment revolution has undeniably transformed how we consume information and entertainment, merging these two seemingly distinct realms into a unified experience. In this final chapter, we will unveil the infotainment epilogue, providing a comprehensive summary of the major themes and findings discussed in the preceding chapters and offering a glimpse into the future of this dynamic industry.

The infotainment landscape has evolved rapidly over the past few decades, driven by advancements in technology and the ever-changing preferences of consumers. From the early days of radio and television broadcasting to the rise of the internet and social media platforms, infotainment has consistently adapted to the needs and desires of its audience. This adaptability has allowed infotainment to permeate nearly every aspect of our lives, from the way we learn about current events to the manner in which we interact with our favorite celebrities and brands.

Throughout this book, we have delved into the various facets of the infotainment industry, examining its history, development, and impact on society. We have also explored the diverse range of infotainment formats and platforms, including news, television, film, music, gaming, and social media. By understanding the intricacies of these different mediums, we have gained a deeper appreciation for the complex and multifaceted nature of infotainment.

In this concluding chapter, we will revisit the major themes and findings that have emerged from our exploration of the infotainment landscape, highlighting the implications and significance of these insights for modern society. We will also address the limitations and critiques of the infotainment industry, acknowledging the challenges and concerns that have arisen as a result of its rapid growth and expansion. Finally, we will offer our final thoughts and recommendations for the future of infotainment, charting a path forward that embraces innovation and ensures the continued evolution of this captivating field.

A Comprehensive Recap of the Infotainment Landscape

As we reach the conclusion of our exploration into the world of infotainment, it is essential to take a step back and reflect on the major themes and findings that have emerged throughout this book. By doing so, we can better understand the infotainment landscape and its implications on modern society.

One of the most prominent themes that has surfaced is the convergence of information and entertainment. In today's fast-paced digital age, consumers are increasingly seeking content that is not only informative but also engaging and entertaining. This has led to the rise of various infotainment formats, such as edutainment, docudramas, and news satire shows, which blend factual information with creative storytelling and humor. This convergence has transformed how we consume information and challenged traditional boundaries between journalism, education, and entertainment.

Another key finding is the role of technology in driving the infotainment revolution. Advancements in digital media, streaming platforms, and social networking sites have made it easier than ever for content creators to produce and distribute infotainment to a global audience. These platforms have also enabled greater interactivity and personalization, allowing users to engage with content in new and immersive ways. As a result, the infotainment industry has experienced rapid growth and diversification, with an ever-expanding array of content catering to diverse interests and demographics.

The democratization of content creation and distribution is another significant theme that has emerged. With the proliferation of affordable and user-friendly digital tools, individuals and small-scale creators now have the ability to produce high-quality infotainment content without the backing of major media corporations. This has led to a more diverse and inclusive infotainment landscape, with a wider range of voices and perspectives being represented.

Throughout this book, we have also examined the various ways in which infotainment can be both beneficial and detrimental to society. On the one hand, infotainment has the potential to educate, inform, and inspire, making complex topics more accessible and engaging for a wider audience. On the other hand, the blurring of lines between fact and fiction, as well as the prioritization of entertainment over accuracy, can lead to misinformation and a superficial understanding of important issues.

Finally, the ethical considerations surrounding infotainment have been a recurring theme. As content creators and consumers, we must be mindful of the potential consequences of our actions and strive to strike a balance between entertainment and responsible information dissemination. This includes being transparent about the sources of information, acknowledging biases, and promoting media literacy among audiences.

In summary, the infotainment landscape is a complex and ever-evolving phenomenon, shaped by the convergence of information and entertainment, technological advancements, and the democratization of content creation. As we continue to navigate this dynamic terrain, we must remain aware of the potential benefits and pitfalls of infotainment and our ethical responsibilities as creators and consumers.

The Far-reaching Effects of Infotainment on Modern Society

The infotainment revolution has undeniably transformed the way we consume information and entertainment, leaving a lasting impact on various aspects of modern society. This section delves into the implications and significance of infotainment, exploring its far-reaching effects on our daily lives, culture, and global communication.

One of the most significant implications of infotainment is the democratization of information. With the advent of the internet and social media, infotainment has become more accessible to the masses, breaking down barriers that once limited the flow of information. This has led to a more informed and engaged public, as people from all walks of life can now access news, entertainment, and educational content at their fingertips.

Infotainment has also played a crucial role in shaping public opinion and discourse. As the line between news and entertainment blurs, audiences are increasingly drawn to content that is both informative and engaging. This has led to the rise of opinion-driven journalism, where news outlets and content creators cater to the interests and biases of their target audience. While this approach has undoubtedly increased audience engagement, it has also raised concerns about the potential for echo chambers and the erosion of objective journalism.

Another significant effect of infotainment is its influence on cultural globalization. As infotainment content transcends geographical boundaries, it fosters a shared global culture that transcends national borders. This has led to the proliferation of international collaborations, cross-cultural exchanges, and the spread of ideas, values, and beliefs across the world. However, this cultural convergence has also sparked debates about cultural homogenization and the potential loss of local and indigenous cultures.

The infotainment revolution has also profoundly impacted the economy, particularly in the media and entertainment industries. The demand for engaging and informative content has led to the growth of new media platforms, job opportunities, and revenue streams. This has also spurred innovation in content creation, distribution, and marketing, as companies and individuals strive to stay ahead in the competitive infotainment landscape.

Despite its numerous benefits, the infotainment revolution has also raised concerns about the quality and credibility of information. The rise of sensationalism, clickbait, and misinformation has led to a growing distrust of media sources, making it increasingly difficult for audiences to discern fact from fiction. This has significant implications for public discourse and the functioning of democratic societies that rely on an informed and discerning citizenry.

In conclusion, the far-reaching effects of infotainment on modern society are both profound and multifaceted. While the democratization of information, cultural globalization, and economic growth are undeniably positive outcomes, the challenges of sensationalism, echo chambers, and misinformation cannot be ignored. As we continue to navigate the infotainment revolution, we must strike a balance between engagement and credibility, ensuring that the content we consume is both entertaining and informative while also fostering critical thinking and discernment among audiences.

Addressing the Challenges and Concerns in the Infotainment Industry

As we have explored the vast landscape of infotainment throughout this book, it is essential to acknowledge that the industry is not without its limitations and critiques. In this section, we will address some of the most pressing challenges and concerns that have emerged as infotainment continues to evolve and reshape our modern society.

One of the primary concerns in the infotainment industry is the blurring of lines between news and entertainment. With the increasing demand for engaging and captivating content, there is a risk that the integrity of journalism may be compromised. News organizations may prioritize sensationalism and clickbait over accuracy and objectivity, leading to a decline in the quality of information available to the public. This phenomenon has given rise to the term "infotainment," which critics argue is a dangerous mix of information and entertainment that can mislead and manipulate audiences.

Another challenge faced by the infotainment industry is the potential for echo chambers and filter bubbles. As consumers increasingly rely on personalized content feeds and social media algorithms, they may become trapped in a self-reinforcing cycle of information that confirms their existing beliefs and biases. This can lead to a fragmented and polarized society, where individuals are less likely to engage in constructive dialogue and debate.

Moreover, the infotainment industry has been criticized for its role in perpetuating stereotypes and reinforcing harmful cultural norms. By prioritizing entertainment value, content creators may inadvertently perpetuate harmful stereotypes and marginalize minority voices. This can have a detrimental impact on society, as it perpetuates ignorance and intolerance.

The commercialization of infotainment also raises concerns about the influence of advertisers and corporate interests on content. As media organizations increasingly rely on advertising revenue, there is a risk that editorial decisions may be influenced by the need to appease sponsors and generate profits. This can lead to a decline in the diversity and quality of content available to audiences and a potential erosion of journalistic independence.

In light of these challenges and concerns, it is crucial for the infotainment industry to engage in self-reflection and strive for continuous improvement. Content creators, journalists, and media organizations must strike a delicate balance between providing engaging content and upholding the principles of accuracy, objectivity, and diversity. By acknowledging and addressing these limitations and critiques, the infotainment industry can continue to evolve and innovate while maintaining its commitment to informing and entertaining audiences responsibly and ethically.

Charting the Path Forward for Infotainment Innovation

As we conclude our exploration into the world of infotainment, it is essential to reflect on the insights gained and consider the future trajectory of this dynamic industry. The infotainment revolution has undoubtedly transformed the way we consume information and entertainment, creating a seamless blend of the two that caters to our ever-evolving digital lifestyles. However, as with any innovation, there are challenges and concerns that must be addressed to ensure the continued growth and success of the infotainment industry. In this final section, we will outline some recommendations for charting the path forward for infotainment innovation.

First and foremost, infotainment creators and providers must prioritize the quality and accuracy of the content they produce. In the age of misinformation and fake news, the responsibility of delivering reliable and trustworthy information cannot be overstated. Infotainment providers must invest in rigorous fact-checking and verification processes to maintain their credibility and uphold the integrity of the industry. Furthermore, striking the right balance between entertainment and information is essential to ensure the audience remains engaged while gaining valuable knowledge.

Secondly, embracing diversity and inclusivity in the infotainment industry is vital for its continued growth and relevance. As the world becomes increasingly interconnected, it is essential to represent a wide range of perspectives, cultures, and experiences in the content we consume. Infotainment providers must actively seek out and promote diverse voices, both in front of and behind the camera, to create a more inclusive and representative media landscape. This fosters a greater sense of empathy and understanding among audiences and enriches the overall quality of the content.

Another recommendation for the infotainment industry is to harness the power of emerging technologies to enhance the user experience. Virtual reality, augmented reality, and artificial intelligence offer exciting possibilities for creating immersive and interactive infotainment experiences that captivate audiences and deepen their engagement with the content. By staying at the forefront of technological advancements, infotainment providers can continue to innovate and push the boundaries of what is possible in this rapidly evolving field.

Moreover, the infotainment industry needs to prioritize ethical considerations and social responsibility in their content creation and dissemination. This includes addressing issues such as privacy concerns, data security, and the potential for manipulation or exploitation of vulnerable audiences. Infotainment providers must establish and adhere to robust ethical guidelines that protect the interests of their users and promote a safe and responsible media environment.

Lastly, fostering collaboration and partnerships within the infotainment industry can help drive innovation and ensure its long-term success. By working together, content creators, technology providers, and other stakeholders can pool their resources, expertise, and ideas to develop groundbreaking infotainment experiences that cater to the diverse needs and preferences of audiences worldwide.

In conclusion, the infotainment revolution has irrevocably changed the way we consume information and entertainment, offering exciting new possibilities for engaging and educating audiences. By addressing the challenges and concerns accompanying this innovation and embracing the recommendations outlined above, the infotainment industry can continue to thrive and evolve, enriching our lives and shaping the future of media consumption.

ABOUT THE AUTHOR

Kiannah Frost is a prolific author and thought leader in the fields of psychology, technology, and social dynamics. With a keen eye for understanding the complexities of human behavior, she has dedicated her career to exploring the ways in which people interact with one another, both in person and through various forms of media.

Kiannah's first series, Mindful Momentum, delved into the power of mindfulness and its impact on personal growth and relationships. Her second series, Breaking Stereotypes, challenged societal norms and encouraged readers to embrace their unique identities. Now, with her third series, Navigating the Digital Age, Kiannah turns her attention to the rapidly evolving world of technology and its influence on human connection.

When she's not writing or coaching, Kiannah enjoys spending time in nature, practicing yoga, and traveling the world, always seeking out new experiences and insights to share with her readers.

Made in the USA
Monee, IL
04 November 2026

67383029R00157